Lecture Notes in Artificial Intelligence 695

Subseries of Lecture Notes in Computer Science
Edited by J. Siekmann

Lecture Notes in Computer Science
Edited by G. Goos and J. Hartmanis

E.P. Klement, W. Slany (Eds.)

Fuzzy Logic
in Artificial Intelligence

8th Austrian Artificial Intelligence Conference,
FLAI'93
Linz, Austria, June 28-30, 1993
Proceedings

Springer-Verlag
Berlin Heidelberg New York
London Paris Tokyo
Hong Kong Barcelona
Budapest

E. P. Klement   W. Slany   (Eds.)

Fuzzy Logic in Artificial Intelligence

8th Austrian Artificial Intelligence Conference,
FLAI '93
Linz, Austria, June 28-30, 1993
Proceedings

Springer-Verlag

Berlin Heidelberg New York
London Paris Tokyo
Hong Kong Barcelona
Budapest

Series Editor

Jörg Siekmann
University of Saarland
German Research Center for Artificial Intelligence (DFKI)
Stuhlsatzenhausweg 3, D-66123 Saarbrücken 11, FRG

Volume Editors

Erich P. Klement
Fuzzy Logic Laboratory Linz
Department of Mathematics
Johannes Kepler University
A-4040 Linz, Austria

Wolfgang Slany
Christian Doppler Laboratory for Expert Systems
Information Systems Department (E184-2)
Technical University of Vienna
Paniglgasse 16, A-1040 Vienna, Austria

CR Subject Classification (1991): I.2, J.2, J.3

ISBN 3-540-56920-0 Springer-Verlag Berlin Heidelberg New York
ISBN 0-387-56920-0 Springer-Verlag New York Berlin Heidelberg

Typesetting: Camera ready by author
Printing and binding: Druckhaus Beltz, Hemsbach/Bergstr.
45/3140-543210 - Printed on acid-free paper

Preface

The Eighth Austrian Artificial Intelligence Conference took place at the Bildungszentrum St. Magdalena, Linz, Austria, June 28-30, 1993. Taking into account the sharply increasing importance of fuzzy logic in many areas of applications, it was decided to focus during this conference on "Fuzzy Logic in Artificial Intelligence".

Out of 34 papers submitted in total, 17 were finally accepted by the international Program Committee for presentation during the conference, all of which are reprinted in this volume. The contributions cover a wide range of areas where fuzzy logic and artificial intelligence meet in current research: theoretical issues, machine learning, expert systems, robotics & control, applications to medicine, and applications to car driving.

In addition to the contributed papers, the conference also featured invited talks by Lotfi A. Zadeh, speaking about "The Role of Fuzzy Logic and Soft Computing in the Conception and Design of Intelligent Systems", and by Irina Ezhkova, speaking about "A Contextual Approach for AI Systems Development". We very much appreciate that both supplied abstracts of their plenary talk for this volume.

Additionally, you will find descriptions of the four workshops that took place during the conference. Johann Gamper and Bernhard Moser organized the "Workshop for Doctoral Students in Fuzzy-Based Systems", Andreas Geyer-Schulz and Peter Kotauczek the one about "Fuzzy Logic for Commercial and Industrial Applications", Rainer Born the one on "Karl Menger, Fuzzy Logic and Artificial Intelligence – An Experiment in Reflection", and Roger Kerr the one on "Fuzzy Scheduling Systems". The conference also included two tutorials, one on "Fuzzy Logic and Applications (in particular to Expert Systems)" by Hans-Jürgen Zimmermann, and one on "Fuzzy Control" by Rudolf Kruse. We are very obliged to all of them for their help in making the conference attractive to all participants.

We are indebted to the members of the Program Committee not only for providing the basis for a fair selection from the initial contributions, but also for many useful comments and suggestions concerning the accepted papers, thus enabling the authors to prepare improved final versions of their contributions.

Finally we would like to thank the supporting companies and institutions and all those persons whose assistance in organizing this conference considerably contributed to its eventual success.

Linz, June 1993   Erich Peter Klement, Wolfgang Slany

Conference Organizer

Österreichische Gesellschaft für Artificial Intelligence /
Austrian Society for Artificial Intelligence

Program Committee

K.-P. Adlassnig (Austria)
M. Fedrizzi (Italy)
V. H. Haase (Austria)
R. Lopez de Mantaras (Spain)
Ph. Smets (Belgium)
R. R. Yager (USA)
H.-J. Zimmermann (Germany)

D. Dubois (France)
A. Geyer-Schulz (Austria/Germany)
R. Kruse (Germany)
M. Reinfrank (Germany)
H. Takagi (Japan/USA)
M. Zemankova (USA)

Additonal Referees

F. Esteva (Spain)
L. Godo (Spain)
H. Hellendoorn (Germany)
N. Honda (USA)
P. S. Khedkar (USA)
R. Palm (Germany)
U. Rehfueß(Germany)

E. Gersthofer (Austria/Germany)
A. Hecht (Germany)
T. Hessberg (USA)
J.-S. Jang (USA)
H. Leufke (Germany)
R. Rehbold (Germany)
M. Reiter (Austria)

Cooperating Institutions

Fuzzy Logic Laboratorium Linz
Christian Doppler Labor für Expertensysteme, Wien

Supporting Companies and Institutions

BEKO - Ing. P. Kotauczek GesmbH., Wien
Technologie und Marketing GesmbH., Linz
European Coordinating Committee for Artificial Intelligence

Table of Contents

Robotics & Control

Applications to Medicine

Applications to Car Driving

Workshop Descriptions

The Role of Fuzzy Logic and Soft Computing in the Conception and Design of Intelligent Systems

Lotfi A. Zadeh

Computer Science Division and the Electronics Research Laboratory,
Department of EECS, University of California, Berkeley, CA 94720;
Telephone: 510-642-4959; Fax: 510-642-5775; E-mail: zadeh@cs.berkeley.edu

Abstract. The past three years have witnessed a significant increase in the rate of growth of MIQ (Machine Intelligence Quotient) of consumer products and industrial systems.

There are many factors which account for the increase in question but the most prominent among them is the rapidly growing use of soft computing and especially fuzzy logic in the conception and design of intelligent systems.

The principal aim of soft computing is to exploit the tolerance for imprecision and uncertainty to achieve tractability, robustness and low solution cost. At this juncture, the principal constituents of soft computing (SC) are fuzzy logic (FL), neural network theory (NN) and probabilistic reasoning (PR), with the latter subsuming genetic algorithms, belief networks, chaotic systems, and parts of learning theory. In the triumvirate of SC, FL is concerned in the main with imprecision, NN with learning and PR with uncertainty. In large measure, FL, NN and PR are complementary rather than competitive. It is becoming increasingly clear that in many cases it is advantageous to employ FL, NN and PR in combination rather than exclusively. A case in point is the growing number of neurofuzzy consumer products and systems which employ a combination of fuzzy logic and neural network techniques.

As one of the principal constituents of soft computing, fuzzy logic is playing a key role in the conception and design of what might be called high MIQ (Machine Intelligence Quotient) systems. There are two concepts within FL which play a central role in its applications. The first is that of a linguistic variable, that is, a variable whose values are words or sentences in a natural or synthetic language. The other is that of a fuzzy if-then rule in which the antecedent and consequent are propositions containing linguistic variables. The essential function served by linguistic variables is that of granulation of variables and their dependencies. In effect, the use of linguistic variables and fuzzy if-then rules results – through granulation – in soft data compression which exploits the tolerance for imprecision and uncertainty. In this respect, fuzzy logic mimics the crucial ability of the human mind to summarize data and focus on decision-relevant information.

A Contextual Approach for AI Systems Development

Irina V. Ezhkova

Russian Academy of Sciences, 40, Vavilova Str.,
Moscow 117333 Russia

Abstract. The method of context formalization based on fuzzy sets theory is suggested. This leads to the development of contextual systems for flexible decision-making in fuzzy environments. Contextual systems have the following features:

- automatic knowledge generation;
- knowledge interpretation and translation from one context to another;
- knowledge adaptation;
- problem solving in a space of contexts.

The example of application (investigation of Sudden Infant Death Syndrome) is presented.

References

1. "Knowledge formation through context formalization." Computers and Artificial Intelligence, vol.8, no.4, 1989.
2. "Contextual Systems for Decision Making in Fuzzy Environments." Proceedings of the Fourth World Congress of International Fuzzy Sets Association (IFSA), Brussels, Belgium, 1991.
3. "Contextual technology for supporting decision making." Cybernetics and systems research'92, edited by R. Trappl, Proceedings of the 11-th European Meeting on Cybernetics and Systems Research, Vienna, vol.1, 1992.
4. "Contextual method for decision making in medicine - case study: Sudden Infant Death Syndrome (SIDS)." Proceedings of the International Conference on Hypoxia and Perinatal Haemodynamics, Turku, Finland, 1992.
5. "Contextual Systems : Is it a way of a Universal Expert System development?" General Systems, New York, 1993 (in print).

Typicality of Concept Instances:
a Semiotic Way for its Evaluation

Anio O. Arigoni

Department of Mathematics ,University of Bologna
P.zza di Porta S. Donato 5, 40127 Bologna, Italy
Tel.: 051-354429 - Fax: 051-354490

Abstract. By this paper we first try to determine the reasons for which Fuzzy Set Theory does not comply satisfactory with the expectations of cognitive psychologists. Then, by utilising results whose achievement was treated in earlier papers, it is reported of an attempt to provide for such purpose. To illustrate the presented development and the achieved results, a detailed paradigmatic example is reported.

1. Introduction

One significant application of Fuzzy Set Theory (FST) was supposed to occur in Cognitive Psychology; specifically, in formalising concepts and operations possible on these. Differently, since the initial tentative in the seventies, cognitivists realised that FST was failing to comply with the requirements expected in the obtainable results. This, notwithstanding: 1rst) it was well established that instances did not have equivalent status in representing concepts (e.g. [23; 22]), so that a fuzzy structure was recognised on these; 2nd) the resolution with which, at that time, FST was indicated as possessing the "natural" tools to assess the *typicality* of concept instances, i. e., the goodness of each instance in representing specific concepts[12; 16; 20; 22].

Consequently, to the early tentative of utilising FST in Cognitive Psychology, the cognitivist community reacted by starting a lively debate. This was risen by Kay [13] and went on by others; among these we mention [15; 23; 24; 31]. One of the main objections regarded the inadequacy of FST to evaluate changes of typicality, which may occur when concepts interact among themselves. About this, some researchers even shown that the application of FST operations to concepts may yield contradictory and/or false results, when applied to concepts [19; 23].

2. Cognitivism and FST

The essential remark that can be made to FST, relatively to its possible utilisation in formalising concepts, is summarised in this section.

Any given concept C can be considered as the knowledge that underlies a category C: the collection, possibly exhaustive, of the elements that are instances of the concept itself. These elements, singly appointed by a name, can so represent the concept itself, alternatively one to another [20]. They may consist in objects, facts or situations and are called *entities*, as in [18].

Treating concepts by FST consists in operating the subsets in which consist the corresponding categories, by the FST connectives; but the results inadequacy can be shown; the following example constitutes a boundary case. Let be given two concepts A and B, the category A relative to the former being a subset of the one B concerning concept B. In Cognitve Psychology, the possibility that for whichever instance x of A, the typicality with respect to B is greater than the one to A itself, as well as the vice versa, may be required [19; 23]. By applying FST to such concepts, i. e. by considering the typicality in topic as a membership function, the exclusion of said possibility is categorical (see for example [14]. This resulted since the first paper on such theory [29] and has been diffused in the literature that followed. It is due to the character of the connectives defined in FST; the membership function of entities resulting from any operation considered on FST depends in fact on the numerical value of the membership the instances themselves have with respect to the operated subsets.

To satisfy the expectations of Cognitive Psychology, the typicality of the instances resulting from operations performed on concepts must depend instead on the context of the resulting subset, exclusively [8]. By the procedure presented in the paper, a contribution to the achievement of the researched solution is illustrated. On the other side, it is observed that also reconsidering the connectives of FST is urged, as for example is arisen from many communications presented at the recent Second IEEE Int. Conf. on Fuzzy Sets (March 29-April 1, 1993, San Francisco, CA.).

The bases on which the approach we present hinges, essentially are: (1) generalisation of categories, a procedure performed to the end of abstracting the concepts underlying these; (2) evaluation of conceptual parameters, like *prototypes* (with respect to the concepts) and entities' *attribute valence*; parameters which, respectivel, are symbolic and numerical in nature [5].

The procedure in subject has roots in the minimisation algorithm of switching functions, introduced by McCluskey [17] and successively considered by Zadeh [30]. We accomplished a first attempt to use such an algorithm for analysing semantic aspects of information [1; 2]; then we applied it to build a mathematical structure deriving from the relations within descriptions of a formal language [3; 4]. Recently the same procedure has been formalised in the framework of Abstraction Theory: *conceptual abstraction* [7]. A theory presented by Plaisted [21] and successively enriched by others researchers [10; 11;12; 25].

The parameters we take into account follows straightway by considering the instances collected to exemplify concepts, as elements of the achieved structures. Thus, an analysis of concepts' articulation in rational thought will be practicable, as for example is looked forward in [9].

The results achieved by delving into the subject-matter dealt with this paper, can be relevant to tackle the development of a concept algebra. This derives from regualrities of rational intelligent processes. In such an algebra operations on concepts can be sensibly carried out and the results these give fulfil the requirements of Cognitive Psychology. This constitutes a complementary extension of the relative theory scientific background which lets so foreseeing new and interesting practical applications, particulary in Machine Learning.

3 Conceptual Abstraction

In the development we present, in order that the concept that underlies a category may be revealed, the elements of the latter undergo to pragmatic generalisation: process performed by means of the semiotic mechanism $\mathcal{M}=\langle \Sigma, R, O \rangle$ here outlined from [7]. In $\mathcal{M}$, Σ is the set of elements through which are described both single entities x_α, which form a set X, and categories C, whose set is $\mathcal{P}(X)$. Such sets form the *simple* and *composite phraseologies*, respectively.

The former, derives from the lexicon consisting in the names —indices h $(=1,...,l)$ — of variables X_h; the field of the latter being $\{0, 1, *\}$. The words these give place to, $x_{h\alpha}$, describe attributes of the entities; they are obtained by one binary interpretation of the values each X_h takes on relatively to each entity. A word tells whether the relative h-th attribute is possessed by the entity itself ("1") or not ("0"), or eventually it is irrelevant ("*"). The same entities result so described by elements x_α of X —ordered l-tuples, with $\alpha=1,...,n$ $(=|X|=3^l)$. These are formed by juxtaposing l words $x_{h\alpha}$; thus, for every α, $x_\alpha = (x_{1\alpha}, ...,x_{l\alpha})$ [3; 4] [1].

The latter, $\mathcal{P}(X)$, is the power set of X deriving from the union of X° and X^*: the set of the elements describing entities all the attribute of which are not relevant and, respectively, that of those in which one or more attributes are such, i.e., are $x_{h\alpha}=*$. Thus $|X^\circ|= 2^l$ and $|X|=|X^\circ \cup X^*|=3^l$. Every composite element describes a category, that is, the entities that separately can represent a specific concept. The simple elements of one same composite element are said to be (one another) *equivalent*.

---

[1] Note that the considered entities could be equivalently represented by conjunctions of *l* ordered ground monoargumental prpredicates of standard first order logic. For computational convenience, we have chosen the 0/1 string representation: those that we simply call *entities*.

The component R of $\mathcal{M}$ consists of relations within the two types of elements, simple and composite. Every element of the power set $\mathcal{P}(X)$ can be derived from others of $\mathcal{P}(X)$ itself, through operations forming the syntax O of $\mathcal{M}$

For simplicity, we will refer to words and to the simple elements that the former give place to, by *attributes* and *entities*, respectively.

3.1 Generalisation of Simple Elements

The possible relation within simple elements regards pairs of entities x_β, x_ξ in one of which, let us say in x_ξ, an attribute, $x_{k\xi}$ is irrelevant, whereas its homologous $x_{k\beta}$ is not such, all other attributes being equal. In this case we say that x_ξ *generalises* x_β, by which x_ξ itself, in turn, is *specified*. This double relation is indicated by $x_\beta > x_\xi$. The latter term, x_ξ results from the conflation of x_β and an x_α pragmatically indistinguishable to it, as described in [25]. The mentioned *indistinguishibility* between x_α and x_β, is meant in the sense of Hobbs [12] and can be formally expressed by

$$(1) \qquad \exists! k: x_{k\alpha} \neq x_{k\beta}, \text{ so that } \forall h \neq k \quad x_{h\alpha} = x_{h\beta}$$

Definitively: *(a)* Given a category C, for every its pair x_a, x_b performing (1) the irrelevance by the k-th attribute to its entities can be asserted and these conflate a third one, x_x, such that:

$$(2) \qquad \forall h \neq k \quad x_{h\xi} = x_{h\alpha} = x_{h\beta} \text{ and } x_{k\xi} = *$$

b) The possible irrelevance of one or more attributes of an x_β, to a concept C is context-depending relatively to all other entities forming the concerning category C.

Every category of entities no one of which includes "*", constitutes a *concrete category*. The subset of the entities of all concrete categories possible in S forms the universal concrete set $X°$ which is a subset of X.

3.2 Abstracting Concepts from Categories

The conflation of all pairs of entities of any concrete category C that perform (1), determines a mapping of these onto the entities of X^*, which form a new category C^1. Although in the entities of the latter some of the attributes that are irrelevant to the concept C underlying C are evident, i.e., even though these being equal to "*" do not correspond to any concrete entity, they still are instances of C itself; simply their description is less specific than the one relative to the entities of C. This means that C can be represented at different levels of specificity, or, conversely, of generality. Thus, between C and C^1 it exists a relation *more-general-than*. This is indicated by ">>", so that it can be written $C >> C^1$.

Through the extensive iteration of conflation on the elements obtained starting from those of C, a sequence of step-by-step more general categories C^j, all relative to C, is obtained. The categories forming such sequence are nested according to $>>$. Formally, for every i,j $(i>j)$ $C^j>>C^i$.

When a category C is exhaustively generalised, another category C^j such that $C^{j+1}=C^j$ is reached: the one within which no more conflations can occur. Concept C still underlies such category; the latter constitutes the most general representation of C, in that in this composite element all the possible attributes irrelevant to C itself are brought into evidence. The category so achieved is called *conceptual abstraction* of C. In this there are detailed the necessary and sufficient conditions for an entity to represent C; it consists in the subset of X —element of $\mathcal{P}(X)$— privileged in representing C and is indicatet by C^A. The subset of X itself representing the same concept by specifying the attributes of all instances of C, is instead formed exclusively by entities of $X°$ and is indicated by C^C.

Finally, by $C^A= \mathcal{F}(C^C)$ is designated the possibility of abstracting the concept C underlying C^C, through the global function $\mathcal{F}$ over $\mathcal{P}(X)$: *conceptual abstraction function*.

3.3 Concept Subsumption

A concept C is said to subsume another, $\mathcal{D}$, when the latter is derived from C itself thropugh the eslusion of some instances of this, according to specific criteria [28]. Thus the cardinality of category D^C gets smaller than that of C^C.

In this case, we say that $\mathcal{D}$ is a sub concept of C.

A synthetic procedure to ascertain whether or not one concept is a sub concept of another, directly from the conceptual abstraction of both, has been analysed [6].

EXAMPLE
A ground-set $X°$ is considered here. Its elements describe entities x_α consisting in persons: the ones that is supposed a hypothetical employment agency is selecting to the end of engagement. The considered attributes of these, $x_{h\alpha}$, are described by words resulting from binary interpretation of the values that four variables, $X_1, ..., X_4$, jointly take on, relatively to each one of such entities. These variables regard:
- X_1 graduate: yes "1", not "0";
- X_2, expected salary: low "0", high "1";
- X_3 knowledge of Japanese : yes "1", not "0";
- X_4, knowledge of Russian: yes "1", not "0".

The ground-set of the different types of person describable through the phraseology generated by the four specific variables, is: $X° = \{x_1=0000, x_2=0001, ..., x_{16}=1111\}$.

It is supposed that a subset of persons that singly can exemplify the concept "interesting" (I), and that preliminary satisfy the one "having college education" (G), is selected on $X°$. Thus the concrete category relative to G is $G^C = \{x_9 = 1000, ..., x_{16} = 1111\}$. The G's *definitional variable* is X_1; this, by assuming value "1" invariably, denotes one attribute that characterises the specific persons having college education. The category relative to I is $I^C = \{x_9 = 1000, x_{10} = 1001, x_{11} = 1010, x_{12} = 1011, x_{14} = 1101, x_{15} = 1110, x_{16} = 1111\}$ (for instance, $x_{11} = 1010$, describes a graduate person, that expects low salary and knows Japanese but does not know Russian). It is supposed that the elements of I^C form one specific subset of G^C, so that G subsumes I, in other terms, the latter is a sub-concept G.

The tabular representation of the same categories G^C and I^C as well as of the relative conceptual abstraction G^A and I^A is reported on Tables I and II.

TABLE I (college education)

G^C		G^A
$x_9 = 1000$		
.	. — *abstraction*	
.	. *process* →	1***
.		
$x_{16} = 1111$		

TABLE II (interesting)

I^C		I^A
$x_9 = 1000$		
$x_{10} = 1001$		
$x_{11} = 1010$ — *abstraction*		10**
$x_{12} = 1011$	*process* →	1**1
$x_{14} = 1101$		1*1*
$x_{15} = 1110$		
$x_{16} = 1111$		

On Table II, concept I is represented, at concrete level, by the entities forming $I^C = \{x_9, x_{10}, x_{11}, x_{12}, x_{14}, x_{15}, x_{16}\}$. It is a description that results bulky and unwanted indeed, because of the number of instances by which is formed; this, even though the considered variables are only four and one of these is merely a definitional one, i. e. a variable denoting constantly one same attribute — either "0" or "1" — in all entities of the concept. The conceptual abstraction of I^C, I^A, instead, consisting of composite elements having a smaller number of generalised entities, allows a reasonably easy represaentation of I: "graduate person that expects a low salary, or that knows Japanese or Russian".

Here we consider three additional concepts by defining them again within the settled $X°$; they are sub-concepts of I and in turn form a sequence of sub-concepts. The cardinality of the categories these underlay decreases, in fact, by the constrains put on their entities, causing so their subsumption one by another. Such concepts are: "convenient to engage (E)", "knowledge of Russian at college level (R)" and "salary strange expectation (S)". Their consistency is shown by the entities listed in Tables III, IV and V, respectively.

TABLE III (convenient)

$$E^C \qquad\qquad E^{\mathcal{A}}$$

$$x_{10}= 1001$$
$$x_{11}= 1010—\ \text{abstraction}$$
$$\qquad\qquad\qquad\qquad\qquad 1**1$$
$$x_{12}= 1011 \qquad\qquad \text{process} \ \rightarrow \ 101*$$
$$x_{14}= 1101$$
$$x_{16}= 1111$$

TABLE IV (Russian)

$$R^C \qquad\qquad R^{\mathcal{A}}$$

$$x_{10}= 1001$$
$$x_{12}= 1011 \quad — \text{abstraction}$$
$$x_{16}= 1111 \qquad\qquad\qquad \text{process} \ \rightarrow 1**1$$
$$x_{14}= 1101$$

TABLE V (strange)

$$S^C \qquad\qquad S^{\mathcal{A}}$$

$$x_{12}= 1011 \quad — \text{abstraction}$$
$$x_{14}= 1101 \qquad\qquad\qquad \text{process} \ \rightarrow \ 1*11$$
$$x_{16}= 1111 \qquad\qquad\qquad\qquad\qquad 11*1$$

4 Category Fuzziness

The gradness of the structure of "category" by now is admitted in Cognitive Psychology as well as in Artificial Intelligence. The problem still open on this topic and which we are dealing with, lies on how to determine this structure; particularly: *(a)* to evaluate its typicality; *(b)* to analyse how the value of this typicality changes when such concept is operated with others, through operations like union, complementation, subsumption and others.

4.1 Instance Validity Preservation

For every concept C, the homologous attributes denoted by one same variable X_h in the entities x_n of the conceptual abstraction $C^{\mathcal{A}}$ and that result as relevant, are identical one another —all these are either "0" or "1". This is well evident from the given example (e.g. the attributes denoted by X_4 in $S^{\mathcal{A}}$). On the other hand, the possible irrelevance of attributes that in the entities of categories less generalised than the conceptual abstraction are different — "1" or "0", respectively — is determined throughout the process of generalisation developed up to obtain $C^{\mathcal{A}}$ as is described in Section 2.3.

It is pointed out that in the concrete representations of the concepts considered in our examples, all the (concrete) entities are explicitly listed; nevertheless the possible irrelevance of their attributes remains hidden. Differently, due to the instances validity preservation and to the efficient complexity reduction accomplished, by "conceptual abstraction":

(1) In the procedure of conceptual abstraction, which entities satisfy the relative concept can be derived through the entities of C^a; e.g. from $x_v=1*11$ of S^a, it is immediate to determine that the (concrete) person of S^c that such entities specify are those described by 1011 and 1111.

(2) By considering one of the concepts, e. g. that I, the attributes that are irrelevant to this, although characterise persons of I^c, they are brought into evidence by considering each x_v of I^a and the x_a of I^c that specify the former (from $x_v=1*1*$ of I^A, it is evident that in $x_{11}=1010$, which is one of the entities that specify it, X_1 and X_3 denote relevant attributes, whereas those denoted by X_2 and X_4 are irrelevant).

From above can be concluded that: the attributes appearing in the entities of any conceptual specification, like those of G^C, I^C,... , which specify the single x_v of G^A, I^A, ... result to be relevant.

Every attribute which is relevant in all entities of a category C^a, is defined as *determinant* to the relative concept C (e.g. the attribute X_4="1" is such to R, as results from the attributes it denotes in the entities x_v of R^c — Table IV —). Any attribute x_{ha} may be determinant to a concept C, and likewise to its opposite, too. In this case the concerning variable X_h is called *discriminant*, to the concept itself (e.g. the attributes connoted by X_1 to the entities of G). If among the variables generating the phraseology X there are one or more that denote irrelevant attributes in all the x_v of a conceptual abstraction C^a as well as in all those of the complement of C^a, then these variables are defined as *absolutely irrelevant*.

It can be proved that the existence of a discriminant variable to a concept C makes all others absolutely irrelevant to the concept itself [3] (examples can be found on Table IV). The concrete representation of the complement of R to G, is the category $G^C \backslash R^C$. For example variable X_4 is determinant by x_{4a} ="0". This attribute is opposite of that x_{4a}="1" which characterises the entities of R. Thus, in the four entities of the considered category, X_2 and X_3 are absolutely irrelevant, whereas X_1 is not such; this in that, being the last a definitional variable, it holds its relevancy in all the x_n of R^a, this it appears in the mentioned table, where such variable is "1" in all entities.

4.2 Concept Prototypes

The idea of concept prototype (then also of that of the concerning category) developed in the context of the paper, is derived from Prototype Theory. According to this, the prototype of a concept must be the "best exemplar" identified by some measure of central tendency [19]. It constitutes the most typical categorisation of the entities forming a category and is designed to represent all the properties of the entities themselves globally [25].

Since it has been proved that in every category C^C the existence and the unicity of an entity such that all its attributes are relevant in one or more x_v of C^a, unless these are absolutely irrelevant [5], we define this entity as *prototype* of this category, as well as of the concept C underlying the latter. Such an entity is indicated by $x_\pi(C)$.

It is noted that for every concept C there may be ascertained the existence also of more than one prototype, this occurs when one or more variables X_h result to be absolutely irrelevant; in this case, however, such prototypes are undifferezntiated, as easily can be realised from the given example; definitively such prototypes pragmatically are all equal among themselves.

Therefore, for the five subconcepts considered in our example the concerning prototypes are:

$x_\pi(G)$ =1*** (a person having college education);

$x_\pi(I)$ =1011 (..., that expects low salary and knows Japanese as well as Russian)

$x_\pi(E)$ =1011 (the same as for I);

$x_\pi(R)$ =1**1 (a person who knows Russian at college level);

$x_\pi(S)$ =1111 (..., who expects high salary and knows Japanese as well as Russian).

4.3 Entity Typicality

The criteria according to which the entities of a category are collected can be thought as based on a sort of *family resemblance*, as suggested by Wittgenstein [27]. Therefore, given a concept C and the category C^C that specifies it, the extent to which every entity x_a of C^c shares its attributes with those characterising $x_\pi(C)$, determines the x_a's effectiveness in representing concept C. Following this reasoning, such a goodness corresponds to the *typicality* sought for, $\tau_C(\alpha)$. This can so constitute the basis to quantify the extensional gradness of every instance to a concept.

Given a concept C, the attributes $x_{h\alpha}$ of its entities x_α, which are equal to those homonymous of the prototype $x_\pi(C)$ —denoted by one same h-th variable— are said to have *(pragmatic) positive valence* to C itself and are indicated by $x_{h\alpha^+}$.

The valence of the attributes opposite to those positive, is *negative*, so that these are indicated by $x_{h\alpha^-}$. By this symbols, entity x_{15}=1110 of category I^C, as an example (see Table II), may be formed by the l-tuple ($x_{1\ 15^+}$, $x_{2\ 15^-}$, $x_{3\ 15^+}$, $x_{4\ 15^-}$).

The entities x_a of the category C relative to a concept C, in which the homologous attributes are $x_{h a^+}$ are a number greater, or at minimum equal, than that of the attributes whose valence is negative. Because of the simplicity we assume for the considered concepts' structure, the homologous attributes of the x_n that form conceptual abstractions are constantly one same; this can be realised, merely by inspection, from G^a, I^a, R^a,... of the given example. From these, for instance, it can be realised that in I, the attributes denoted by variable X_2,, x_{2a}="0" which are positive are four, whereas those negative —x_{2a}="1"— are only three.

The evaluation of the degree of valence of the different attributes is carried out in three steps; in these, the following parameters are considered:

(a) The degree of positive valence that each attribute $x_{h\alpha}$ possesses to a concept C, $\varpi_C(h\alpha)$; it depends on the difference between the number of entities x_α of category C^C in which such an attribute is $x_{h\alpha^+}$ and those in which the same is negative, related to the cardinality of C^C itself. This degree is expressed by:

$$(3) \qquad \varpi_C(h\alpha) = (|\{x_\alpha \in C^C: x_{h\alpha}^+\}| - |\{x_\alpha \in C^C: x_{h\alpha}^-\}|)/|C^C|$$

(b) the relative typicality of each entity x_a of the specific description of C, C^C; it is indicated by $\tau_C(\alpha)°$, and expresses the valence expectation, to l, of the attributes $x_{h a}$ that are identical to their homologous of the prototype x_{hp}, that is, having positive valence; thus:

$$(4) \qquad \tau_C(\alpha)° = \sum x_{h\alpha^+} \varpi_C(h\alpha) / l$$

Finally, (c) for every x_α the *normalised typicality* results by relating $\tau_C(\alpha)°$ to the corresponding value of the prototype, $\tau_C(\pi)°$, as it results from

$$(5) \qquad \tau_C(\alpha) = \tau_C(\alpha)° / \tau_C(\pi)°$$

It is remarked that for every concept, the concerning x_p has the maximum value of typicality $\tau_C(\pi)°$ and its normalisation leads $\tau_C(\pi)$ to unit.

By (4) the values of relative typicality of the prototypes concerning the different concepts S, I, ... of the given example result:

$$\tau_G(\pi)° = .25; \quad \tau_I(\pi)° = .36; \quad \tau_E(\pi)° = .60; \quad \tau_R(\pi)° = .50; \quad \tau_S(\pi)° = .50$$

The relative typicality of the entities forming the five considered concepts is obtained through (3) and (4). Their values are reported on Table VI; they regard single entities, with respect to each concerning sub concepts.

Table VI (entities relative typicality)

	$\tau_S(\alpha)°$	$\tau_R(\alpha)°$	$\tau_E(\alpha)°$	$\tau_I(\alpha)°$	$\tau_G(\alpha)°$
x_9	-	-	-	.28	.25
x_{10}	-	.50	.45	.32	.25
x_{11}	-	-	.35	.32	.25
x_{12}	.58	.50	.50	.36	.25
x_{13}	-	-	-	-	.25
x_{14}	.58	.50	.40	.28	.25
x_{15}	-	-	-	.36	.25
x_{16}	.66	.50	.45	.32	.25

A comparison among the values of typicality of every entity x_α to the sub concepts, S, R..., after their normalisation through (5), is interesting in that it indicates the achievement of our prefixed goal. Particularly significant to the purpose is that within the values relative to entity x_{14}. This (=1101) describes "a graduate person who expects high salary, does not know Japanese but knows Russian" and its resulting dynamics to the different sub concepts S, R, ... , is represented by the diagram reported in Figure 1. Also standing that category S is included in R, this in E, and so forth, the values in topic start from .87, $\tau_S(14)$, to rise to 1, $\tau_R(14)$, there is a drop to .80, $\tau_E(14)$, a permanence at .80, $\tau_I(14)$, then, finally, 1 is reached once again, $\tau_G(14)$;

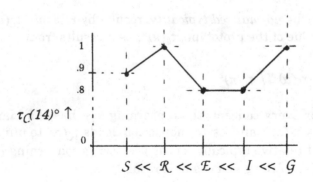

Fig . 1 - Variation of the (normalised) typicality value of x_{14}
with respect to the nested sub concepts S, R, E, I, G.

The typicality of the considered person —entity x_{14}— may changes from one sub concept to another in both senses, either up or down, or eventually remains constant. From the reported diagram, it is evident how the order in which the inclusion between categories relative to the subsumed concepts occurs, does not result directly relevant to the sense of such changes. The typicality by each one of the instancesto to any sub concept, result in fact overall depending on the contextual relevance of the attributes characterising the instances themselves.

5 Conclusions

The necessity of a rigorous semiotic determination of the typicality of the instances to fuzzy concepts has been discussed in the paper. Its evaluation is carried out on the basis of the contextual dependence of each instance of a concept, on the *semantic values* of the other instances of the latter.

By the proposed procedure, the typicality to one concept, results as a kind of a *pragmatic proximity* of an instance to all others pertaining the concept itself, considered globally. This proximity is conferred on every instance by the positive attriutes characterising it, weighed by the degree of valence of this.

The reported development constitutes a preliminary remark to adduce the presentation of operations that makes the relative theory complying with Cognitive Psychology, and that are suggested by the original ideas of fuzzy sets. Thus the obtained typicality can give a contribution to solve problems connected with concept machine learning.

References

[1] A. O. Arigoni: "Structure of the semiotic dimensions of information", *Proceed. of the Int. Congr. for Semiotic Studies*, Milano, 1974.
[2] A. O. Arigoni and E. Balboni:"Definizione ricorsaiva di significato soggettivo", *Proc. of Congr. Naz. di Filosofia della Scienza*, 403-412, L'Aquila, Italy, (1974).
[3] A. O. Arigoni: "Mathematical development arising from 'semanticimplication' and the evaluation of membership characteristic functions", *Fuzzy Sets and Systems*, No.4, 167-183,1980.
[4] A. O. Arigoni: "Transformational-generative grammar for describing formal properties", *Fuzzy Sets and Systems*, 8, 311-322, 1982.
[5] A. O. Arigoni & C. Furlanello: "Typicality in Learning through Conceptual Clustering", *Proceedings of AI*IA - 89*, 1989.
[6] A. O. Arigoni and V. Maniezzo: "Incremental Algebraic Concept Learning", *11th Int. Conf. on Cybernetic and System Research*, Wien, 1992.
[7] A. O. Arigoni & V. Maniezzo: "Conceptual abstraction: a context-depending simplification of concept representations", *Working Notes of the Approximation and Abstraction of Computational Theories '92"*, San Jose, Calif. 1992.
[8] A. O. Arigoni: "Fuzzy structured concepts and typicality of their instances", *Proced. of the Second IEEE Int. Conf. on Fuzzy DSystems*, II, 1309-1315, S. Francisco CA (USA), 1993.
[9] D. Dubois: "Semantic et Cognition: catégories, prototypes, typicalité", Editions du CNRS, Paris, 1991.
[10] A. Giordana, D. Roverso and L. Saitta: "Abstracting background knowledge for concept learning", *Prooc. of Machine learning-EWSL-91*, 1-13, Springer Verlag, 1991.

[11] F. Giunchiglia and T. Walsh:"A theory of Abstractation", *Artificial Intelligence*, 57, 323-389, 1992.

[12] Hobbs, J.R.: "Granularity", in Proceed. of 9th IJCAI Conference, 432-435, *Int. Joint Conf. on Artificial Intelligence*, 1985.

[13] P. Kay: "A model-theoretical approach to folk taxonomy", *Soc Sci. Inf.*, 14, 151-166, 1975.

[14] G. Lakoff: "Hedges: A study in meaning criteria and the logic of Fuzzy Concepts", *Jour. of Phil. Logic*, 2, 458-508, 1973.

[15] G. V. Jone: "Staks not fuzzy sets: An ordinal basis for prototype theory of concepts", *Cognition*, 12, 281-290, 1982.

[16] M. McCloskey and S. Clucksberg "Decisonprocessesin verifying class-inclusion statements: implication for model of semantic memory", *Cognitive Psychology*, 11, 1-37, 1976

[17] E.J. McLuskey: "Introduction to the Theory of Switching Circuits", New York, McGraw-Hill 1965.

[18] R. S.Michalski and R. E. Stepp: "Conceptual clustering of structured objects: a goal-oriented approach", *Artificial Intelligence*, 28, pag.43-69, 1986.

[19] D. N. Oberson and E. E. Smith: "On the adequacy of prototype theory as a theory of concepts", *Cognition*, 9, 52-72, 1981.

[20] G. C. Oden: "Fuzzyness in semantical memory: chosing exmples of subjective categories", *Memory and Cognition*, 5, 198-204, 1977.

[21] D. A. Plaisted: "Theorem proving with abstarction", *Artificial Intelligence*, 16, 47-108, 1981.

[22] E. Rosc: "Cognitive representation of semantic categories", *J. Exp. Psychology: general*, 104, 1975.

[23] E.M. Roth and C. B. Mervis: "Fuzzy set theory and class inclusion relations in semantic categories", *J. of Verbal Learning and Verbal Behaviour*, 22, 509-525, 1983.

[24] P. Tabossi: "La rappresentazione mentale dei concetti", *Giornale Italiano di Psicologia*, 1, 45-70, 1985.

[25] J.D. Tenenberg: "Preserving Consistency Across Abstraction Mappings", *IJCAI-10*, 1987.

[26] A. Vandierendonck: "Are category membership decision based on concept gradness ?", *Eur. Jour. of Cogni Phych.*, 3 (3), 343-362, 1991.

[27] L. Wittgenstein: "Philosohpical Investigation", Oxford, Backwell, 1953.

[28] Was, L. et all, "Automated Reasoning introduction and applications", Prentice-Hall, Inc. New Jersey, 1984.

[29] L. A. Zadeh: "Fuzzy Sets", *Information and Control*, 8, 3, 33-353, 1965.

[30] L. A. Zadeh: "A fuzzy approach to the definition of complex or imprecise concepts", *Int. J. Man-Machine Studies*, 8, 249-291, 1976.

[31] L. A. Zadeh: "A note on prototype theory", *Cognition*, 12, 291-297, 1982.

Non-conventional Conjunctions and Implications in Fuzzy Logic*

János C. Fodor   and   Tibor Keresztfalvi

Eötvös Loránd University, Budapest 112, P. O. Box 157, H-1502 Hungary

Abstract. First, we make some remarks concerning the definition of connectives in fuzzy logic. We point out possible disadvantages of considering exclusively t-norms and t-conorms as proper models for the conjunction and disjunction. Coincidence of S- and R-implications is investigated by solving functional equations for conjunctions. Then, we suggest a constructive approach to axiomatics of the generalized modus ponens (GMP). Besides a special model, a particular class of conjunctions satisfying the axioms for GMP and based on the Hamacher family of t-norms is also characterized. On the other hand, the coincidence of R- and S-implications defined by the members of this class is verified.

1 Introduction

The proper definition of connectives is one of the most important problems of fuzzy logic. It is needless to emphasize the dominance of t-norms, t-conorms, strong negations and related implications. Their sound theoretical foundation as well as their wide variety have given them almost an exclusive role in different investigations and applications. However, people are inclined to use them also as a matter of routine: for example, when one works with binary conjunctions and there is no need to extend them for three or more arguments, associativity is an unnecessarily restrictive condition. The same is valid for commutativity if the two arguments have a different semantical background and it has no sense to replace them with each other. There are also some addititioanal reasons that advocate the study of enlarged classes of binary operations for fuzzy sets:

- Obviously, properties of conjunctions, disjunctions and negations have to be connected and to be in accordance with those of fuzzy implications. However, if one compares usual axioms for fuzzy implications with properties of R- and S-implications defined by t-norms, t-conorms and strong negations, then it can easily be observed that these two families have 'much nicer' properties than it would be axiomatically expected. For more details see Weber [1], Dubois and Prade [2], Fodor [3].
- There is no way to define strict negations via t-norm-based residuation: the resulted negation is either degenerate or strong, see 4.2 Remark and 4.3 Theorem in [1].

---

* Partially supported by OTKA.

- t-norm-based R- and S-implications are, in general, different. They can coincide if and only if the underlying t-norm is isomorphic to the Lukasiewicz
  t-norm, see for instance Smets and Magrez [4].
- In the inference pattern called *generalized modus ponens* (GMP for short),
  a number of intuitively desirable properties are not possessed using t-norms
  and implications defined by t-norms. For more details, see Magrez and Smets
  [5].

These observations, which are very often left out of consideration, have urged
us to revise definitions and properties of operations in fuzzy logic. A new unifying
approach is suggested for investigation of these operations: it is supported by an
important connection between implications and conjunctions given by equation
(4) below.

The paper is organized as follows. After some necessary preliminaries we
state our first problem in Section 2. The resulted functional equation (4) is
solved in Section 3. In Section 4 we revisite the generalized modus ponens,
choosing a constructive way to investigate its properties. This leads us to a
system of functional equations for conjunction and implication in GMP. After
giving a particular solution, we determine conjunctions, besides Lukasiewicz-like
t-norms, for which the issued R- and S-implications coincide and, at the same
time, they provide us appropriate models of connectives in GMP.

2 Definitions and problem setting

In this section we recall some definitions and results that are more or less known
in the literature. Then we introduce our problem that will be investigated and
solved.

A function $n : [0,1] \rightarrow [0,1]$ is called *negation* if it is non-increasing and
$n(0) = 1$, $n(1) = 0$. A negation n is called *strict* if n is continuous and decreasing.
A strict negation n is called *strong* if $n(n(x)) = x$ for every $x \in [0,1]$.

A binary operation $*$ on $[0,1]$ is called a *fuzzy conjunction* if it is an extension
of the classical Boolean conjunction, i.e.,

$$x * y \in [0,1] \quad \text{for every} \quad x, y \in [0,1]$$

and

$$0 * 0 = 0 * 1 = 1 * 0 = 0; \quad 1 * 1 = 1.$$

A canonical model of fuzzy conjunctions is the family of t-norms, i.e., functions $T : [0,1] \times [0,1] \rightarrow [0,1]$ which are commutative, associative, non-decreasing
and $T(x,1) = x$ for every $x \in [0,1]$. For more details see e.g. Weber [1] and
Schweizer and Sklar [7].

A binary operation $\rightarrow$ on $[0,1]$ is a *fuzzy implication* if it is an extension of
the Boolean implication, i.e.,

$$x \rightarrow y \in [0,1] \quad \text{for every} \quad x, y \in [0,1]$$

and

$$0 \to 0 = 0 \to 1 = 1 \to 1 = 1; \quad 1 \to 0 = 0.$$

Let $\odot$ be any binary operation on $[0,1]$. The following transformations of $\odot$ play a central role in this paper:

$$x\mathcal{S}_n(\odot)y = n(x \odot n(y)), \tag{1}$$
$$x\mathcal{R}(\odot)y = \sup\{z \in [0,1] \mid x \odot z \le y\}, \tag{2}$$

where n is a strong negation.

Obviously, $\mathcal{S}_n \circ \mathcal{S}_n(\odot) = \odot$ for any binary operation $\odot$ on $[0,1]$ (here $\circ$ denotes composition). Moreover, $\mathcal{S}_n(\to)$ is a fuzzy conjunction if $\to$ is a fuzzy implication. On the other hand, if $*$ is a fuzzy conjunction then $\mathcal{S}_n(*)$ and $\mathcal{R}(*)$ are fuzzy implications. It is clear that

$$I_S(x, y) = x\mathcal{S}_n(*)y \qquad \text{S-implication}$$

is based on the classical view of implications while

$$I_R(x, y) = x\mathcal{R}(*)y \qquad \text{R-implication}$$

is based on a residuation concept, see e.g. Dubois and Prade [8] when $*$ is a t-norm and Fodor [3] when $*$ is an arbitrary fuzzy conjunction.

One can define a sequence of conjunctions $\{*_j\}$ in the following way ($*_0 = *$ is a conjunction):

$$*_j = \mathcal{S}_n \circ \mathcal{R}(*_{j-1}), \quad j = 1, 2, 3, \ldots. \tag{3}$$

Our first problem can be stated as follows: characterize those conjunctions $*$ for which the above sequence (3) is closed, i.e., for which there exists a positive integer m such that

$$*_m = *. \tag{4}$$

The solution is given in the next section.

3 Closure theorems

All results of this section (with more details and proofs) can be found in Fodor [3, 9]. Fortunately, it is sufficient to investigate the above problem for $m = 1$ and $m = 2$ due to the following theorem.

Theorem 1 *Let $\{*_j\}$ be a sequence of conjunctions defined by (3). Then*
  (a) $*_{2k+1} = *$ *holds for some $k \ge 0$ if and only if $*_1 = *$.*
  (b) $*_{2k} = *$ *is valid for some $k > 0$ if and only if $*_2 = *$.*

It is clear from the definition of $\{*_j\}$ that $*_1 = *$ is equivalent to

$$\mathcal{R}(*) = \mathcal{S}_n(*) \tag{5}$$

while $*_2 = *$ means that

$$\mathcal{R} \circ \mathcal{S}_n \circ \mathcal{R}(*) = \mathcal{S}_n(*). \tag{6}$$

Moreover, (5) implies (6).

The situation described by the second equation, which was investigated by Dubois and Prade [8] in the case when $*$ is a t-norm and by Fodor [3] in the general case, is illustrated on Figure 1.

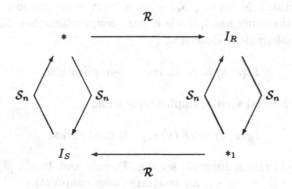

Fig. 1. The second case

Complete characterizations of binary operations satisfying either equation (5) or equation (6) are given in the following theorem.

Theorem 2 *A binary operation $*$ on $[0,1]$ satisfies equation*
  (a) $\mathcal{R}(*) = \mathcal{S}_n(*)$ *if and only if*

$$x * z \leq y \quad \Longleftrightarrow \quad x * n(y) \leq n(z) \quad \forall x, y, z \in [0,1];$$

  (b) $\mathcal{R} \circ \mathcal{S}_n \circ \mathcal{R}(*) = \mathcal{S}_n(*)$ *if and only if*

$$x * z \leq y \quad \Longleftrightarrow \quad z \leq x \mathcal{R}(*) y \quad \forall x, y, z \in [0,1].$$

It is worth drawing up the corresponding results when $* = T$ is a t-norm.

Corollary 1 *A t-norm T, as a binary operation on $[0,1]$, fulfils*
  (a) equation (5) if and only if there exists an automorphism φ of the closed unit interval such that

$$T(x,y) = \varphi^{-1}(\max\{\varphi(x) + \varphi(y) - 1, 0\})$$

and

$$n(x) = \varphi^{-1}(1 - \varphi(x));$$

(b) equation (6) if and only if T is left-continuous in both places on $(0, 1]$.

In other words, t-norm-based R- and S-implications coincide if and only if T is a φ-transform of the Lukasiewicz t-norm. Another class of conjunctions, for which (5) also holds, will be characterized in Section 5. This class of conjunctions satisfies some properties which makes it suitable for using in approximate reasoning, especially in the generalized modus ponens.

4 The generalized modus ponens

The *generalized modus ponens* (GMP), an inference pattern with fuzzy predicates, is given as follows:

Rule	if S_1 has property A then S_2 has property B	
Fact	S_1 has property A'	(7)
	S_2 has property B'	

where A, A' and B, B' are fuzzy sets of the universes X and Y respectively, i.e. $A, A' \in \mathcal{F}(X)$ and $B, B' \in \mathcal{F}(Y)$. We emphasize that these fuzzy sets are not necessarily normalized.

B' is calculated as:

$$B'(y) = \sup_x M\big(A'(x), I_{A \to B}(x, y)\big), \qquad (8)$$

where M is a fuzzy conjunction and $I_{A \to B}$ is a fuzzy binary relation (usually an implication) on $X \times Y$.

Usually, GMP is expected to meet a number of intuitively desirable requirements. Most papers on GMP investigate this problem by choosing first particular classes of conjunctions (e.g. t-norms) and implications (e.g. S- or R-implications based on t-norms) then testing whether the different requirements are fulfilled. There are lots of possible choices but still no *best* one, see [5].

Opposed to these approaches, we choose a constructive way to investigate properties of GMP. We regard M and $I_{A \to B}$ as operations defined in common sense, see Section 2. Then, we fix only a few basic requirements to be fulfilled, in our opinion, by GMP. They lead to a system of functional equations for M and $I_{A \to B}$. In order to find a solution we assume some reasonable properties of conjunction and implication operators. Then a particular solution for M and $I_{A \to B}$ is given and some further properties of GMP are verified as consequences, however they usually appear as requirements in the rich literature on GMP (see e.g. references in [5]).

Notice that a different approach, a new model of fuzzy modus ponens, was established also in [5] in order to satisfy all the intuitively required properties. Instead, we keep GMP unchanged while conjunctions and implications are used in a broad sense.

4.1 Axioms

First, we assume that $I_{A \to B}$ is defined pointwise, that is,

A1 $I_{A \to B}(x, y)$ depends only on $A(x)$ and $B(y)$ i.e. $I_{A \to B}(x, y) = J(A(x), B(y))$

and so (8) turns into

$$B'(y) = \sup_x M\big(A'(x), J(A(x), B(y))\big) \tag{9}$$

In the literature it is generally required that

R1 if $A' = A$ then $B' = B$ ($A, B \not\equiv 0$);
R2 if $\operatorname{Supp} A' \cap \operatorname{Supp} A = \emptyset$ then $B' \equiv 1$ ($A, B \not\equiv 0$);
R3 monotonicity: $B'(y)$ is non-decreasing with respect to $A'(x)$ and $B(y)$ and non-increasing with respect to $A(x)$;
R4 if $A' \equiv 0$ then $B' \equiv 0$.

R1 reflects the coincidence of (9) with classical modus ponens. R2 forces the GMP to infer *unknown* when the fact A' has nothing to do with the antecedent A. R3 is clear and R4 is also obvious: if nothing is observed then nothing is inferred. Note that neither commutativity nor associativity was considered above.

First, we want to find at least one pair (M, J) such that R1–R4 are satisfied by using (9).

4.2 Conditions on M and J implied by the crisp case

Obviously, the GMP should satisfy the above properties also when A, A', B, B' are crisp sets, so we obtain from (9) on the basis of R1, R2 and R3 that

$$\max \big\{ M(0, J(0, 1)), M(1, J(1, 1)) \big\} = 1 \tag{10}$$

$$\max \big\{ M(0, J(0, 0)), M(1, J(1, 0)) \big\} = 0 \tag{11}$$

$$\max \big\{ M(0, J(1, v)), M(1, J(0, v)) \big\} = 1 \quad (v \in \{0, 1\}) \tag{12}$$

After simple calculations we finally get from the above equations and from R3 and R4 the following system of equations for any $u, v \in [0, 1]$

$$
\begin{aligned}
M(0, J(u, v)) &= 0 \\
M(1, J(0, v)) &= 1 \\
M(u, J(1, 0)) &= 0 \\
M(1, J(u, 1)) &= 1
\end{aligned}
\tag{13}
$$

Replacing A, A' and B by fuzzy singletons (fuzzy points) of height u and v respectively, we have from R1 for any $u, v \in]0, 1]$ the following equation:

$$M(u, J(u, v)) = v \tag{14}$$

In order to solve the system (13)-(14) we need some further reasonable assumptions:

A2 M is non-decreasing with respect to both arguments (shortly $M(\nearrow, \nearrow)$);

A3 $M(0, v) = 0 \quad \forall v \in [0, 1]$;

A4 $M(u, v) \leq v \quad \forall u, v \in [0, 1]$;

A5 J is non-increasing with respect to its first argument and non-decreasing with respect to its second argument (shortly $J(\searrow, \nearrow)$);

A6 $J(0, v) = 1 \quad \forall v \in [0, 1]$;

A7 $J(1, v) \leq v \quad \forall v \in [0, 1]$.

By using these assumptions it is easy to see that

$$M(1, v) = v$$
$$J(1, v) = v \tag{15}$$
$$J(u, 1) = 1$$

Then (14) and (15) together imply that

$$M(u, 1) = 1 \quad \forall u > 0. \tag{16}$$

Compare equation (14) and properties A2, A3, A4, (16) with those of a modus ponens generating function in Trillas and Valverde [10].

4.3 A solution and some consequences

A possible solution for M and J satisfying all the assumptions and the system (13)-(14)-(15) is given by

$$M(u, v) = \begin{cases} \dfrac{u}{u + 1 - v} & \text{if } 0 < u \leq v \\ v & \text{if } u > v \\ 0 & \text{if } u = 0 \end{cases} \tag{17}$$

$$J(u, v) = \begin{cases} 1 + u - u/v & \text{if } u \leq v, \quad v \neq 0 \\ v & \text{if } u > v \\ 1 & \text{if } u = v = 0 \end{cases} \tag{18}$$

It is worth mentioning that J is just the R-implication generated by M and fits the general framework in Section 2. Furthermore M is idempotent, which is useful in dealing with redundancies in knowledge bases, see [2].

Using this pair of M and J the following properties are also satisfied by (9):

P1 if $A' \subset A$ then $B' \subset B$;

P2 if $A' \equiv 1$ and $\exists x : A(x) = 0$ then $B' \equiv 1$;

P3 if $A \equiv 0$ and $A' \not\equiv 0$ then $B' \equiv 1$;

P4 if $A \equiv 1$ and $\operatorname{hgt} A' \geq \operatorname{hgt} B$ then $B' = B$.

Finding all solutions (M, J) of (13)-(14)-(15) is a subject of further research. A particular class of solutions for which R- and S-implications (resp. R- and S-conjunctions) coincide are given in the following section.

5   A particular class of connectives

Having a look at equautions (15) and (16) it seems to be reasonable to search
for appropriate new operations (both for conjunctions and implications) in the
following form:
$$\frac{T(x,y)}{x},$$
where $x \in (0,1]$ and $y \in [0,1]$ and T is a t-norm. The choice of this form was
also motivated by a formula in [8]:
$$\max\left(\frac{x+y-1}{x},0\right).$$

Thus, assume that T is a t-norm. Define a new binary operation on $(0,1] \times [0,1]$ by
$$H(x,y) = \frac{T(x,y)}{x}. \qquad (19)$$
Operation H has the following basic properties, for any t-norm T:

 - $H(x,y) \in [0,1]$ for any $(x,y) \in (0,1] \times [0,1]$;
 - H is non-decreasing with respect to its second argument but, in general,
   nothing can be said about the first one;
 - $H(x,1) = 1$, $H(x,0) = 0$ for any $x \in (0,1]$;
 - $H(1,y) = y$ for any $y \in [0,1]$;

We introduce operation H^C by
$$H^C(x,y) := \begin{cases} H(x,y) & \text{if } x > 0 \\ 0 & \text{otherwise} \end{cases} \qquad (20)$$
if H is non-decreasing with respect to its both arguments and operation H^I by
$$H^I(x,y) := \begin{cases} H(x,y) & \text{if } x > 0 \\ 1 & \text{otherwise} \end{cases} \qquad (21)$$
if H is non-increasing with respect to its first argument and non-decreasing with
respect to the second one.

Then H^C is a fuzzy conjunction and H^I is a fuzzy implication in the broad
sense of Section 2.

We can define S- and R-implications based on H^C in the usual way, using
the standard strong negation $n(x) = 1 - x$:
$$J_S(x,y) = 1 - H^C(x, 1-y), \qquad (22)$$
$$J_R(x,y) = \sup\{z|H^C(x,z) \le y\}, \qquad (23)$$
and similarly S- and R-conjunctions based on H^I by
$$M_S(x,y) = 1 - H^I(x, 1-y), \qquad (24)$$
$$M_R(x,y) = \inf\{z|H^I(x,z) \ge y\}. \qquad (25)$$

We have seen in Section 3 that no other t-norms but those, isomorphic to Lukasiewicz's one in sense of Corollary 1, satisfy the closure condition (5). Now, we look for operations of type (19) satisfying this condition, that is, we would like to characterize continuous t-norms T for which (22) and (23) or respectively (24) and (25) coincide, see Figure 2.

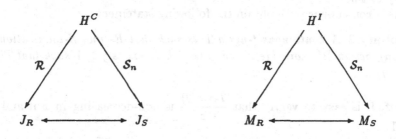

Fig. 2. Coincidence of R- and S-transforms

5.1 Conjunctions for which R- and S-implications coincide

Observe first that $J_S(x, y) = J_R(x, y)$ implies the following equation for T:

$$T\left(x, 1 - \frac{T(x, 1-y)}{x}\right) = xy \qquad \forall (x, y) \in (0, 1] \times [0, 1]. \tag{26}$$

Indeed, let us recall a well-known fact: if T is a continuous t-norm and I_T is the R-implication defined from T by residuation, i.e.

$$I_T(x, y) := \sup\{z | T(x, z) \le y\},$$

then for any $x, y \in [0, 1]$ we have

$$T(x, I_T(x, y)) = \min\{x, y\}. \tag{27}$$

Using now the simple relationship

$$J_R(x, y) = I_T(x, xy) \qquad \forall x, y \in [0, 1]$$

the equation $J_S(x, y) = J_R(x, y)$ can be written as

$$1 - H^C(x, 1 - y) = I_T(x, xy),$$

and taking into account (27) we immediately obtain for any $(x, y) \in (0, 1] \times [0, 1]$ that

$$T\left(x, 1 - \frac{T(x, 1-y)}{x}\right) = T(x, I_T(x, xy)) = \min\{x, xy\} = xy.$$

Equation (26) was completely solved by Fodor and Keresztfalvi [11]. The result is given in the next theorem.

Theorem 3 *A continuous t-norm T satisfies functional equation (26) if and only if there exists a $\gamma \geq 0$ such that*

$$T(x, y) = T_\gamma(x, y) := \frac{xy}{\gamma + (1 - \gamma)(x + y - xy)} \tag{28}$$

This theorem means that equation (26) is a characterization of the Hamacher family of t-norms.

As a consequence, we obtain the following statement.

Corollary 2 *A continuous t-norm T is such that R- and S-implications based on conjunction H^C coincide if and only if there exists $\gamma \geq 1$ such that $T(x, y) = T_\gamma(x, y)$.*

Proof. It is easy to veryfy that $\dfrac{T_\gamma(x, y)}{x}$ is non-decreasing in x if and only if $\gamma \geq 1$. ∎

Closing this section, we present the parameterized form of conjunction H^C as follows ($\gamma \geq 1$):

$$H^C(x, y) := \begin{cases} \dfrac{y}{\gamma + (1 - \gamma)(x + y - xy)} & \text{if } x > 0 \\ 0 & \text{otherwise} \end{cases} \tag{29}$$

It is easy to check that the pair (H^C, J_S) fulfils equation (14) for all $\gamma \geq 1$, where

$$J_S(x, y) = 1 - H^C(x, 1 - y) = \begin{cases} \dfrac{\gamma y + (1 - \gamma)xy}{\gamma + (1 - \gamma)(1 - y + xy)} & \text{if } x > 0 \\ 1 & \text{otherwise} \end{cases}, \tag{30}$$

i.e. this pair is a suitable model of conjunction and implication in GMP for all $\gamma \geq 1$.

5.2 Implications for which R- and S-conjunctions coincide

In this subsection we mention shortly the corresponding problem of implications. Namely, we characterize those t-norms for which S- and R-conjunctions (24) and (25) coincide. Now our basic equation is

$$1 - H^I(x, 1 - y) = \inf\{z | H^I(x, z) \geq y\}. \tag{31}$$

As in the previous subsection, equality $M_S(x, y) = M_R(x, y)$ implies also functional equation (26) for T. By the general result in Theorem 3, we can easily prove the following statement.

Corollary 3 *A continuous t-norm T is such that R- and S-conjunctions based on implication H^I coincide if and only if there exists $0 \leq \gamma \leq 1$ such that $T(x, y) = T_\gamma(x, y)$.*

It is worth mentioning that $M_S(x, y) = 1 - H^I(x, 1 - y)$ with H^I satisfy equation (14). That is, connectives (M_S, H^I) are well fitted for the generalized modus ponens.

6 Conclusion

In this paper we investigated fuzzy conjunctions and implications from different points of view. By the results it became clear that one must be rather flexible in choosing connectives for particular reasons. Especially, non-commutative and non-associative conjunctions such as (29) and the corresponding implications (30) can fulfil the expected properties better than t-norms and related implications. Therefore, we would like to encourage readers to use more advanced operators like these or (17) and (18) not only in theoretical problems but also in practice.

References

1. Weber, S.: A general concept of fuzzy connectives, negations and implications based on t-norms and t-conorms. Fuzzy Sets and Systems 11 (1983) 115–134
2. Dubois, D., and Prade, H.: Fuzzy sets in approximate reasoning, Part 1: Inference with possibility distributions. Fuzzy Sets and Systems 40 (1991) 143–202
3. Fodor, J. C.: On fuzzy implication operators. Fuzzy Sets and Systems 42 (1991) 293–300
4. Smets, P., and Magrez, P.: Implications in fuzzy logic. Int. J. Approximate Reasoning 1 (1987) 327–347
5. Magrez, P., and Smets, P.: Fuzzy modus ponens: a new model suitable for applications in knowledge-based systems. Internat. J. Intelligent Systems 4 (1989) 181–200
6. Hamacher, H.: Über logische Verknüpfungen unscharfer Aussagen und deren zugehörige Bewertungsfunktionen. Working Paper 75/14, RWTH Aachen, Aachen, Germany (1975)
7. Schweizer, B., and Sklar, A.: Probabilistic Metric Spaces. North-Holland, Amsterdam, 1983
8. Dubois, D., and Prade, H.: A theorem on implication functions defined from triangular norms. Stochastica VIII (1984) 267–279
9. Fodor, J. C.: A new look at fuzzy connectives. Technical Report TR 91/1, Computer Center, Eötvös L. University, Budapest, Hungary, 1991.
10. Trillas, E., and Valverde, L.: On mode and implication in approximate reasoning. In: M. M. Gupta, A. Kandel, W. Bandler, J. B. Kiszka (eds.): Approximate Reasoning in Expert Systems. North-Holland, Amsterdam, 1985, pp. 157–166
11. Fodor, J. C., and Keresztfalvi, T.: Non-standard connectives in fuzzy logic. Proc. of 2nd IEEE Int. Conf. on Fuzzy Systems, San Francisco, March 28 - April 1, 1993

A comparative fuzzy modal logic

Petr Hájek and Dagmar Harmancová

Institute of Computer Science, Academy of Sciences
182 07 Prague, Czech Republic

Abstract. A formal logical systems dealing both with uncertainty (possibility) and vagueness (fuzziness) is investigated. It is many-valued and modal. The system is related to a many-valued tense logic. A completeness theorem is exhibited.

1 Introduction

The distinction between vagueness (fuzziness) and uncertainty has been stressed by several authors. The former concerns degrees of truth and leads to many-valued logics, whereas the latter concerns degrees of belief and is best formalized by some sort of modal logic. Cf. e.g. [Godo]. The main difference concerns presence or absence of truth-functionality. Fuzzy logic deals with fuzzy propositions that may have intermediate degrees of truth (related often to values of some quantity like height etc.) and is usually understood as truth-functional, i.e. the truth-degree of a compound formula (conjunction, disjunction, negation, etc.) is a function of the truth degrees of its components (computed using generalized truth tables). On the other hand, uncertainty as degree of belief into the truth of a crisp proposition is best understood as some measure (not necessarily probabilistic measure) on the set of all possible worlds (possible states, elementary events) assigning to a proposition p the measure of the set of all worlds in which p is true - and as such it is not truth functional. We shall prefer possibility measures; and a possibility measure Π satisfies $\Pi(A \vee B) = max(\Pi(A), \Pi(B))$ but the possibility of $A\&B$ is not a function of $\Pi(A), \Pi(B)$. This remembers modal logic with modalities $\Diamond$ (possibly) and $\Box$ (necessarily). For most modal systems, $\Diamond(A \vee B)$ is equivalent to $\Diamond A \vee \Diamond B$ but $\Diamond(A\&B)$ is not equivalent to $\Diamond A\&\Diamond B$. The mathematical framework to define semantics of modal logic - Kripke models - generalizes often to systems with other modalities.

Even if formulas bear possibilities, which are numerical values, we may be interested not in the values themselves but in their comparison, i.e. investigate formulas $A \lhd B$ saying that B is at least as possible as A. Here $\lhd$ behaves as a modality generalizing in some sense the modality $\Diamond$ (possibly). The logical language using $\lhd$ is qualitative (or we can say: comparative), i.e. does not have means to express possibilities as numbers but only their comparisons. This leads to interesting, well-defined and natural logical systems and the question naturally emerges, if they are related to some classical systems of modal logic. In [Bendová-Hájek] a qualitative possibilistic modal logic is studied and related to a tense (temporal) logic with finite linearly preordered time. Here we analyze a logical system dealing both with fuzziness and uncertainty, which means that it is both many-valued and modal. Many-valued modal logics have been studied by [Fitting]. Our system is denoted by QFL^*; it is a qualitative finitely-valued fuzzy possibilistic logic. Building such a system we immediately meet a problem: how to compare possibilities of fuzzy formulas? It turns out that such an entity is best understood as a fuzzy truth value, i.e. as a fuzzy subset of truth values. Surprisingly or not, we have several choices of comparing possibilities of fuzzy formulas; we shall offer two and choose one of them. Our system is related to a finitely valued tense logic with linearly preordered time (denoted by MTL^*); we get a completeness theorem for MTL^* and a theorem on a faithful interpretation of QFL^* in MTL^*. Several problems are formulated.

2 A qualitative fuzzy logic

We fix $n + 1 \geq 2$ - the number of truth values. *Values* is the set of truth values. Our *symbols* are propositional variables, connectives $\land, \lor, \rightarrow, -$ and (i) for each $i = 0, 1/n, 2/n, ..., 1$, as well a modality $\lhd$.

Formulas are built in the obvious way, (i) being unary (if A is a formula then so is $(i)A$); the modality $\lhd$ is binary (if A, B are formulas then $A \lhd B$ is a formula). Example of a formula: $(A \land -(1)A) \lhd A$.

We use the following *truth-tables*:

$\delta_\land(i, j) = min(i, j)$, $\delta_\lor(i, j) = max(i, j)$, $\delta_\rightarrow$ is Gödel's implication, $\delta_{(i)}(j) = 1$ if $j = i$ and $= 0$ otherwise; $\delta_-(x) = 1 - x$ (usual fuzzy negation). Note that Gödel's negation $\neg$ is definable: $\neg A$ is $A \rightarrow (0)A$.

A fuzzy possibilistic (Kripke) structure is a structure $K = \langle W, \Vdash, \pi \rangle$ where $W \neq \emptyset$, $\Vdash$ maps *Atoms* $\times W$ into *Values* (thus each atom has a truth value in each world) and π is a normalized fuzzy subset of W - a possibility distribution ($\pi(w)$ is the possibility of w for $w \in W$; if $X \subseteq W$ then $\pi(X)$ is defined as $sup_{w \in X} \pi(w)$). We write $\| p \|_w$ instead of $\Vdash(p, w)$ and extend $\Vdash$ to all Boolean combinations of atoms using truth tables; thus e.g. $\| A \land B \|_w = min(\| A \|_w, \| B \|_w)$, $\| (i)A \|_w = 1$ iff $\| A \|_w = i$, $\| (i)A \|_w = 0$ otherwise, etc. To define the semantics of the modality $\lhd$ we use fuzzy truth values.

Two fuzzy truth values are given by each formula A, $\tau_{\pi,A}$ and $\sigma_{\pi,A}$ defined as follows:

$$\tau_{\pi,A}(z) = sup\{\pi(w) \mid \| A \|_w = z\},$$

$$\sigma_{\pi,A}(z) = sup\{\pi(w) \mid \| A \|_w \geq z\}.$$

Obviously, σ is non-increasing. We shall base our semantics for $\lhd$ on σ, not on τ. The main reason is the validity of a very natural deduction rule, see Fact (1) below. The modality $A \lhd B$ should satisfy the following:

$$\| A \lhd B \|_w = 1 \quad \text{iff} \quad (\forall z)(\sigma_{\pi,A}(z) \leq \sigma_{\pi,B}(z))$$

(read for example:"B is at least as possible as A"). The question, what is $\|A \lhd B\|_w$ if the defining condition does not hold, is postponed.

We shall investigate 1-truth, i.e. formulas such that $\| A \|_w = 1$ for all w. Extend $\Vdash$ to all modal formulas, i.e. nested modalities are allowed.

Lemma.

$$\sigma_{\pi, A \vee B}(z) = \max(\sigma_{\pi, A}(z), \sigma_{\pi, B}(z)),$$

$$\sigma_{\pi, A \wedge B}(z) \leq \min(\sigma_{\pi, A}(z), \sigma_{\pi, B}(z)).$$

Facts. (1) The rule "from $A \to B$ infer $A \lhd B$" is sound for 1-truth (and Gödel's implication, possibly for others too).

(2) The following are 1-tautologies:

$$A \lhd A,$$

$$((A \lhd B) \wedge (B \lhd C)) \to (A \lhd C),$$

$$\text{false} \lhd A \lhd \text{true},$$

$$((A \lhd C) \wedge (B \lhd C)) \to (A \vee B) \lhd C,$$

Definition. The comparative structure given by a Kripke model K is $Comp(K) = \langle W, \Vdash, \leq \rangle$, where $w \leq w'$ iff $\pi(w) \leq \pi(w')$.

Thus the comparative structure results if the possibility π (a real-valued function) is replaced by the induced linear preorder. This resembles tense logics (logics of time); their model have the form $\langle W, \Vdash, \leq \rangle$ where $\leq$ is some (linear or partial) order. We shall exhibit a generalized many valued tense logic closely related to our fuzzy possibilistic logic.

3 A many valued tense logic

The logic in question, denoted by MTL^*, will have the same modality-free formulas, truth values and truth tables as the logic QFL^* described in the previous section; but it will have three unary modalities G, H, I (read "in all future worlds, in all past worlds, in all present worlds" respectively).

Kripke models have the form $\langle W, \Vdash, \leq \rangle$, where $\leq$ is a linear preorder on W. The semantics of the modality G is as follows:

$$\| GA \|_w = \inf_{w' > w} \|A\|_{w'}$$

analogously H ($\inf_{w' < w}$), I ($\inf_{w' \equiv w}$). We further define $FA \equiv -G(-A)$, $PA \equiv -H(-A)$, $JA \equiv -I(-A)$, $\Box A \equiv GA \wedge IA \wedge HA$, $\Diamond A \equiv -\Box(-A)$, $\hat{G}A \equiv GA \wedge IA$, $\hat{F}A \equiv -\hat{G}(-A)$ etc. Formulas $FA, PA, JA, \hat{G}A, \hat{H}A, \hat{F}A, \hat{P}A, \Box A, \Diamond A$ are read: A holds in some future world, some past world, some present world; in all future and present worlds, in all past and present worlds, in some future or present worlds, in some past or present worlds, in all worlds, in some worlds. Finally we define

$$A \lhd^* B \equiv \Box(A \to \hat{F}B);$$

thus $\|A \lhd^* B\| = \inf \{\| \hat{F}B \|_w \mid \| A \|_w > \| \hat{F}B \|_w\}$.

We can take this for the definition of $\| A \lhd B \|_w$ in QFL^*. By this the definition of the semantics of QFL^* has been completed. We continue to define MTL^* and our aim is to get a completeness result for MTL^*.

A formula is called *Boolean* if it results from formulas of the forms $(i)A$ using connectives and modalities; clearly, if B is Boolean then $\| B \|_w$ is 0 or 1.

Axioms. (a) **Propositional axioms:** some choice of axioms complete for the given propositional calculus, cf. [Gottwald]: some few axioms for implication and other connectives; furthermore,

(1) $\bigvee_i (i)A$,

(2) $\bigwedge_{i \neq j} \neg((i)A \wedge (j)A)$,

(saying that each formula has exactly one truth value),

(3) $(i)A \wedge (j)B \to (min(i,j))(A \wedge B)$

and similarly for other connectives $\vee, \to, -$, e.g.

(3') $(i)A \wedge (j)B \to (1)(A \to B)$ if $i \leq j$ etc.

(3'') $(1)A \to A$ for arbitrary A and also $A \equiv (1)A$ for A Boolean

(3''') $A \vee -A$ for A Boolean.

(b) Modal axioms.

(4) $G(A \to B) \to (GA \to GB)$,

$GA \to GGA$

$PGA \to A, A \to GPA$

similarly for H, F, I, J;

(5) $F(A) \wedge F(B) \to (F(A \wedge FB) \vee F((FA) \wedge B) \vee F(A \wedge IB))$

(linearity, analogously for P);

(6) $IA \to A$

$JA \to IJA$ (S5 for I, J);

(7) $G(\geq i)A \equiv (\geq i)GA$

$G(\leq i)A \equiv (\leq i)FA$,

analogously for H, P, I, J

(Note that $(\geq i)A$ is $\bigvee_{j \geq i}(j)A$ etc.)

Deduction rules of this modal logic: modus ponens, necessitation for G, H, I (e.g. "from A infer GA"), and also "from A infer $(1)A$".

Lemma. All these are 1-sound for $(n+1)$-valued Kripke models with linearly preordered time.

Remark. Axioms (4)-(5) are Burgess's $(A0) - (A2)$, i.e. his L_2, modified only in (5) by inserting I (preorder); (4)-(6) corresponds also to modal axioms of FLPOT [Bendová-Hájek]. Axioms (7) are from [Fitting].

Completeness theorem. $MTL^* \vdash A$ (A is provable in our tense logic) iff A has the value 1 in all worlds of all Kripke models with linearly preordered time.

The *proof* is a variant of [Burgess]'s proof of completeness of the (two-valued) tense logic with linearly ordered time and is presented in Sect.4. The proof in fact gives the following

Lemma. If a formula A is not provable in MTL^* then there is a Kripke structure $\langle W, \Vdash, \leq \rangle$ with W at most countable and such that for some $w \in W$, $(< 1)A$ is true in w.

Corollary. (1) A formula A of MTL^* is a 1-tautology (with respect to all models $\langle W, \Vdash, \leq \rangle$) iff A is 1-true in all models $Comp(K)$ where $K = \langle W, \Vdash, \pi \rangle$ is a model of QFL^*.

(2) Consequently, the interpretation of QFL^* in MTL^*, associating with each formula A of QFL^* the formula A^* of MTL^* resulting by replacing $\lhd$ by $\lhd^*$, is faithful with respect to 1-tautologies. (Caution: do not overlook that models of QFL^* bear a possibility distribution on possible worlds, whereas models of MTL^* bear a linear preorder on them.) The next section is denoted to a proof of the last lemma.

Remark. (1) The problem remains to find an axiomatization of 1-tautologies of QFL^* consisting of some formulas of QFL^* itself.

(2) The task remains to consider the relation of QFL^* to the many valued tense logic with finite linearly preordered time, in an analogy to [Bendová-Hájek].

(3) Moreover, one can investigate logics based on comparison of other fuzzy truth values, infinitely valued systems and many other variations. The purpose of the present paper is mainly to show the direction of future research.

4 Proof of completeness

Lemma. The following are provable formulas:

(11) $(1)A \wedge (1)(A \rightarrow B)) \rightarrow (1)B$

(12) $(FA\&GB) \rightarrow F(A\&B)$

(13) $F(\geq i)A \equiv (\geq i)FA$ for $i > 0$;
$\quad F(\geq 0)A \equiv F(true)$

(14) $F(\leq i)A \equiv (\leq i)GA$ for $i < 1$;
$\quad F(\leq 1)A \equiv F(true)$

(15) $(i)GA \equiv G(\geq i)A \wedge F(i)A$ for $i < 1$
$\quad (1)GA \equiv G(1)A$

(16) $(i)FA \equiv G(\leq i)A \wedge F(i)A$ for $i > 0$

(17) $(i)IA \rightarrow I(i)IA$

Similarly for corresponding other modalities (for I, J no restrictions to i are necessary).

Proof hints. (13) and (14) follow from (7); (15) follows also from (7) using $\vdash (i)GA \equiv (\geq i)GA \wedge (\leq i)GA$ and $\vdash F(\leq i)A\&G(\geq iA) \to F((i)A)$ (by (12)). The proof of (16) is similar. To prove (17) use the I-variant of (15), i.e. $\vdash (i)IA \equiv I(\geq i)A\&J(i)A$ (arbitrary (i)); but $\vdash I(\geq i)A \to II(\geq i)A$ and $\vdash J(i)A \to IJ(i)A$, thus $\vdash (i)IA \to [II(\geq i)A\&IJ(i)I(A)] \to I[I(\geq i)A\&J(i)IA] \to I(i)IA$.

Definition. A *theory* is a set of Boolean formulas, containing all provable Boolean formulas (in particular, T contains (1)A for each provable Boolean formula A). A Boolean formula is *provable in* T if it has a propositional proof from members of T using only modus ponens. T is *inconsistent* if $T \vdash A$ and $T \vdash -A$ for some Boolean A. T is *complete* if for each Boolean A, $T \vdash A$ or $T \vdash -A$.

Fact. Each consistent theory T has a consistent complete extension. (Proof obvious thanks to Ax.(3''').)

Fact. If $T \supseteq \{FA\}$ is consistent (A being boolean) then a $T' \supseteq \{A\}$ is also consistent.

Proof: If $B_1, ... B_n$ is a propositional proof of a contradiction from A, then each FB_i is provable from FA. This makes FA inconsistent.

Definition. Let T be a complete consistent theory; for each formula A, put $e(A) = i$ iff T contains the formula (i)A. An *evaluation* is an e given by a complete theory T.

Lemma. If e is as above then e commutes with connectives, i.e. $e(A \wedge B) = min(e(A), e(B))$ etc.; furthermore, $e(A) = 1$ for each logically provable A.

Proof. We have $e(A \wedge B)$ iff $(i)(A \wedge B) \in T$; also the following formula is in T: $(j)A \wedge (k)B \to (min(j, k))(A \wedge B)$. Take j, k such that $j = e(A), k = e(B)$; then $(j)A \wedge (k)B \in T$, hence $(min(j, k))(A \wedge B) \in T$, thus $i = min(j, k)$.

Lemma Let e, e' be evaluations. Then the following are equivalent:
(1) for each A, $e(GA) \leq e'(A)$
(2) for each A, $e(A) \leq e'(PA)$
(3) for each A, $e(A) \geq e'(HA)$
(4) for each A, $e(FA) \geq e'(A)$

Proof. Assume (1) and prove (2): since $A \to GPA$ is an axiom, $e(A) \leq e(GPA) \leq e'(PA)$.

Assume (1) and prove (4): let $e'(A) = i$. Then $e'(-A) = 1 - i$, $e(G(-A)) \leq 1 - i$, $e(FA) = e(-G(-A)) = 1 - e(G(-A)) \geq i = e'(A)$.

The rest is similar.

Definition. For evaluations e, e', let $R(e, e')$ mean that for each A, $e(GA) \leq e'(A)$. Furthermore, let $E(e, e')$ mean that for each A, $e(IA) = e'(IA)$.

Lemma. Let $e_0 R e_1$, $e_0 R e_2$ (e_i evaluations). Then $e_1 R e_2$ or $e_2 R e_1$ or $e_1 E e_2$.

Definition. A *chronicle* is a system $C = (W, (e_x)_{x \in W}, \leq)$ when $W \neq \emptyset$, each e_x is an evaluations, $\leq$ is a linear preorder and, for each $x, y \in W$,

$$x < y \text{ implies } e_x R e_y,$$

$$x \equiv y \text{ implies } e_x E e_y.$$

C is *perfect* if for each x, A,

$e_x(FA) > 0$ implies that for some $y, z > x$, $e_x(FA) = e_y(A)$, and $e_x(GA) = e_z(A)$ (this condition is denoted by (a, x, A)),

$e_x(PA) > 0$ implies that for some $y, z < x$, $e_x(PA) = e_y(A)$ and $e_x(HA) = e_z(A)$ (denoted (b, x, A)),

$e_x(JA) > 0$ implies that for some $y, z \equiv x$, $e_x(JA) = e_y(A)$ and $e_x(IA) = e_z(A)$ (denoted (c, x, A)).

$C' = (W', (e_x)_{x \in W'}, \leq')$ is an *extension* of C if $W \subseteq W'$ and $\leq$ is $(\leq') \cap W^2$.

Killing lemma. If C is a finite chronicle and x, A are such that (a, x, A) is violated then there is a finite extension C' of C in which (a, x, A) holds; all conditions satisfied in C remain satisfied in C'.

Sublemma. If $e(FA) > 0$ then there are e', e'' such that eRe', eRe'' and $e'(A) = e(FA)$, $e''(A) = GA$.

Proof. Let $e(FA) = i$; we show that the theory $T' = \{(i)A\} \cup \{(\geq j_B)B \mid e(GB) = j_B\}$ is consistent. Any completion of T' defines an e'. Take $B_1, ..., B_k$ and let $j_m = e(GB_m)$. Let T be the theory defining e.

$$T \vdash (i)FA\& \bigwedge_m (j_m)GB_m,$$

thus

$$T \vdash F((i)A\& \bigwedge_m (\geq j_m)B_m),$$

thus $\{(i)A\} \cup \{(\geq j_m)B_m \mid m\}$ is consistent. By compactnes, T' is consistent.

Now assume $i = e(GA) > 0$ and let T' be as above (with the new i). Then $T \vdash (i)GA$, thus $T \vdash (\leq i)GA$, $T \vdash F(\leq i)A$, but we also have $T \vdash (\geq i)GA$, thus $T \vdash G(\geq i)A$, which gives $F(i)A$. Then continue as above.

Now assume $i = e(JA) > 0$, let $j = e(IB)$ (put $k = 1$ for brevity).

$$T \vdash (i)JA\&(j)IB$$

$$T \vdash (i)JA\&I(j)IB$$

$$T \vdash J((i)A\&(j)IB).$$

Finally, assume $i = e(IA) > 0$,

$$T \vdash (i)IA\&(j)IB,$$

$$T \vdash I(\geq i)A\&J(i)A\&I(j)IB,$$

$$T \vdash J((i)A\&(j)IB).$$

This completes the proof of the sublemma.

Proof of the lemma. Let $j = e_x(FA) > 0$, assume that there is no $y > x$ such that $e_y(A) = j$. Take an e' such that $e_x Re'$, $e'(A) = j$ and extend W by adding an y such that $e_y = e'$. The only question is how to extend the ordering. You may assume that x is the greatest element such that $e_x(FA) = j$ (otherwise replace x by this greatest element). If x has no successor, just add y as the new greatest element. If x has an immediate successor x' then observe that

$e_{x'}(FA) < j$, thus the relation $e_{x'}Re_y$ does not hold, hence either $e_{x'}Ee_y$ or $e_y Re_{x'}$. Thus either we make y equivalent to x' or put y between x and x'.

Similarly for other cases, the case (c, x, A) being simpler. This proves the lemma. By an obvious countable iteration we may satisfy all the requirements and get the following.

Lemma. Each finite chronicle can be extended to a finite or countable perfect chronicle.

Last lemma. If $C = \langle W, (e_x)_w, \leq \rangle$ is a prefect chronicle and we define $K = \langle W, \Vdash, \leq \rangle$ by putting $\| p \|_x = e_x(p)$ for each variable p then for each x, A,

(*) $\qquad \| A \|_x = e_x(A)$.

Proof. By induction on the complexity of A. Connectives are obvious except $(i)A$. Assume (*); $e_x((i)A) = 1$ iff $e_x(A) = i$, otherwise $e_x((i)A) = 0$ thus we have the induction step. Modalities can be handled using

$$(i)GA \equiv G(\geq i)A \& F(i)A \quad (\text{for } i < 1)$$

$$(1)GA \equiv G(1)A$$

Indeed, assume $i < 1$ and (*). By definition, $\| GA \|_x = min_{x'>x} \| A \|_{x'}$; but if $e_x(GA) = i$ then for each $x' > x$ we have xRx', i.e. $e_{x'}(A) \geq i$; and if $i < 1$ then $e_x((i)FA) = 1$ and by killing, there is an $x'' > x$ such that $e_{x''}(A) = i$. Thus $e_x(GA) = min_{x'>x}e_{x'}(A)$, which gives $\| GA \|_x = e_x(GA)$. Similarly for other modalities.

Completeness theorem. $\vdash A$ (A is provable in our logic) iff A has the value 1 in all worlds of all Kripke models with linearly preordered time.

Proof. If A is unprovable then $(1)A$ is unprovable, thus $(< 1)A$ is a consistent Boolean formula and has a model K such that in a world w of K $(< 1)A$ is true, i.e. $\| A \|_w < 1$.

References

1. Bendová K.,Hájek P.: Possibilistic logic as tense logic, in: Proc. of QUARDET'93, Barcelona.

2. Burgess J.P.: Basic tense logic, in: Gabbay, Guenthner (eds.): Handbook of Philosophical Logic, Vol. II, Reidel 1984.

3. Fitting M.: Many-Valued Modal Logics, Fundamenta Informaticae.

4. Godo L., Lopez de Mantaras R.: Fuzzy logic, to appear in Encyclopaedia of Computer Science.

5. Gottwald S.: Mehrwertige Logik, Akademie-Verlag, Berlin 1988

Combining Neural Networks and Fuzzy Controllers

Detlef Nauck, Frank Klawonn, and Rudolf Kruse

Dept. of Computer Science, Technical University of Braunschweig
Bueltenweg 74 - 75, D-38106 Braunschweig, Germany

Abstract. Fuzzy controllers are designed to work with knowledge in
the form of linguistic control rules. But the translation of these linguistic rules into the framework of fuzzy set theory depends on the choice
of certain parameters, for which no formal method is known. The optimization of these parameters can be carried out by neural networks,
which are designed to learn from training data, but which are in general
not able to profit from structural knowledge. In this paper we discuss
approaches which combine fuzzy controllers and neural networks, and
present our own hybrid architecture where principles from fuzzy control
theory and from neural networks are integrated into one system.

1 Introduction

Classical control theory is based on mathematical models that describe the behavior of the plant under consideration. The main idea of fuzzy control [11, 14],
which has proved to be a very successful method [7], is to build a model of a
human control expert who is capable of controlling the plant without thinking
in a mathematical model. The control expert specifies his control actions in the
form of linguistic rules. These control rules are translated into the framework
of fuzzy set theory providing a calculus which can simulate the behavior of the
control expert.

The specification of good linguistic rules depends on the knowledge of the
control expert. But the translation of these rules into fuzzy set theory is not formalized and arbitrary choices concerning for example the shape of membership
functions have to be made. The quality of the fuzzy controller can be drastically influenced by changing shapes of membership functions. Thus methods for
tuning fuzzy controllers are necessary.

Neural networks offer the possibility to solve this tuning problem. The combination of neural networks and fuzzy controllers assembles the advantages of
both approaches and avoids the drawbacks of them. Although a neural networks
is able to learn from given data, the trained neural network is generally understood as a black box. Neither is it possible to extract structural knowledge from
the trained neural network, nor can we integrate special information about the
problem into the neural network in order to simplify the learning procedure. On
the other hand, a fuzzy controller is designed to work with knowledge in the
form of rules. But there exists no formal framework for the choice of parameters
and the optimization of parameters has to be done by hand.

In Sections 2 and 3 we discuss approaches which use ordinary neural networks to optimize certain parameters of an ordinary fuzzy controller, and describe how neural networks can preprocess data for a fuzzy controller and extract fuzzy control rules from data. These approaches strictly separate the tasks of the neural network from the task of the fuzzy controller. Section 4 and 5 are devoted to hybrid architectures integrating the concepts of a fuzzy controller into a neural network or understanding a fuzzy controller as a neural network.

2 Tuning Fuzzy Control Parameters by Neural Networks

For the implementation of a fuzzy controller it is necessary to define membership functions representing the linguistic terms of the linguistic inference rules, which is more or less arbitrary. For example, consider the linguistic term *approximately zero*. Obviously, the corresponding fuzzy set should be a unimodal function reaching its maximum at the value zero. Neither the shape, which could be triangular or Gaussian, nor the range, i.e. the support of the membership function is uniquely determined by *approximately zero*. Generally, a control expert has some idea about the range of the membership function, but he would not be able to argue about small changes of his specified range.

Therefore, the tuning of membership functions becomes an import issue in fuzzy control. Since this tuning task can be viewed as an optimization problem neural networks offer a possibility to solve this problem.

A straightforward approach is to assume a certain shape for the membership functions which depends on different parameters that can be learned by a neural network. This idea was carried out in [16] where the membership functions are assumed to be symmetrical triangular functions depending on two parameters, one of them determining where the function reaches its maximum, the other giving the width of the support. Gaussian membership functions were used in [8].

Both approaches require a set of training data in the form of correct input–output tuples and a specification of the rules including a preliminary definition of the corresponding membership functions.

A method which can cope with arbitrary membership functions for the input variables is proposed in [6, 19, 20]. The training data have to be divided into r disjoint clusters $R_1, \ldots, R_r$. Each cluster R_i corresponds to a control rule $\mathcal{R}_i$. Elements of the clusters are tuples of input–output values of the form (x, y) where x can be a vector $x = (x_1, \ldots, x_n)$ of n input variables.

This means that the rules are not specified in terms of linguistic variables, but in the form of crisp input–output tuples.

A multilayer perceptron with n input units, some hidden layers, and r output units can be used to learn these clusters. The input data for this learning task are the input vectors of all clusters, i.e. the set $\{x \mid \exists i \, \exists y : (x, y) \in R_i\}$. The target output $t_{u_i}(x)$ for input x at output unit u_i is defined as

$$t_{u_i}(x) = \begin{cases} 1 & \text{if there exists } y \text{ s.t. } (x, y) \in R_i \\ 0 & \text{otherwise.} \end{cases}$$

After the network has learned its weights, arbitrary values for x can be taken as inputs. Then the output at output unit u_i can be interpreted as the degree to which x matches the antecedent of rule $\mathcal{R}_i$, i.e. the function

$$x \mapsto o_{u_i}$$

is the membership function for the fuzzy set representing the linguistic term on the left–hand side of rule $\mathcal{R}_i$.

In case of a Mamdani type fuzzy controller the same technique can be applied to the output variable, resulting in a neural network which determines the fuzzy sets for the right–hand sides of the rules.

For a Sugeno type fuzzy controller, where each rule yields a crisp output value together with a number, specifying the matching degree for the antecedent of the rule, another technique can be applied. For each rule $\mathcal{R}_i$ a neural network is trained with the input–output tuples of the set R_i. Thus these r neural networks determine the crisp output values for the rules $\mathcal{R}_1, \dots, \mathcal{R}_r$.

These neural networks can also be used to eliminate unnecessary input variables in the input vector x for the rules $\mathcal{R}_1, \dots, \mathcal{R}_r$ by neglecting one input variable in one of the rules and comparing the control result with the one, when the variable is not neglected. If the performance of the controller is not influenced by neglecting input variable x_j in rule $\mathcal{R}_i$, x_j is unnecessary for $\mathcal{R}_i$ and can be left out.

3   Neural Networks for Extracting Fuzzy Rules

In the previous section we described, how neural networks could be used to optimize certain parameters of a fuzzy controller. We assumed that the control rules where already specified in linguistic form or as a crisp clustering of a set of correct input–output tuples. If we are given such a set of input–output tuples we can try to extract fuzzy (control) rules from this set. This can either be done by fuzzy clustering methods [3] or by using neural networks.

The input vectors of the input–output tuples can be taken as inputs for a Kohonen self–organizing map, which can be interpreted in terms of linguistic variables [17]. The main idea for this interpretation is to refrain from the winner–take–all principle after the weights for the self–organizing map are learned. Thus each output unit corresponds to an antecedent of a fuzzy control rule. Depending on how far output unit u_i is from being the 'winner' given input vector x, a matching degree $\mu_i(x)$ can be specified, yielding the degree to which x satisfies the antecedent of the corresponding rule.

Finally, in order to obtain a Sugeno type controller, to each rule (output unit) a crisp control output value has to be associated. Following the idea of the Sugeno type controller, we could choose the value

$$\frac{\sum_{(x,y) \in S} \sum_i \mu_i(x) \cdot y}{\sum_{(x,y) \in S} \mu_i(x)}$$

where S is the set of known input–output tuples for the controller.

Another way to obtain directly a fuzzy clustering is to apply the modified Kohonen network proposed in [4].

Kosko uses another approach to generate fuzzy–if–then rules from existing data [10]. Kosko shows that fuzzy sets can be viewed as points in a multidimensional unit hypercube. This makes it possible to use *fuzzy associative memories (FAM)* to represent fuzzy rules. Special adaptive clustering algorithms allow to learn these representations (AFAM).

4 Fuzzy Control Oriented Neural Networks

Another approach to combine fuzzy control and neural networks is to create a special architecture out of standard feed–forward nets that can be interpreted as a fuzzy controller. In [1] such a system, the ARIC architecture (*approximate reasoning based intelligent control*) is described by Berenji. It is a neural network model of a fuzzy controller and learns by updating its prediction of the physical system's behavior and fine tunes a predefined control knowledge base.

This kind of architecture allows to combine the advantages of neural networks and fuzzy controllers. The system is able to learn, and the knowledge used within the system has the form of fuzzy–if–then rules. By predefining these rules the system has not to learn from scratch, so it learns faster than a standard neural control system.

ARIC consists of two special feed–forward neural networks, the *action–state evaluation network (AEN)* and the *action selection network (ASN)*. The ASN is a multilayer neural network representation of a fuzzy controller. In fact, it consists of two separated nets, where the first one is the fuzzy inference part and the second one is a neural network that calculates a *measure of confidence associated with the fuzzy inference value* using the weights of time t and the system state of time $t + 1$. A *stochastic modifier* combines the recommended control value of the fuzzy inference part and the so called "probability" value and determines the final output value of the ASN.

The hidden units of the fuzzy inference network represent the fuzzy rules, the input units the rule antecedents, and the output unit represents the control action, that is the defuzzified combination of the conclusions of all rules (output of hidden units). In the input layer the system state variables are fuzzified. The membership values of the antecedents of a rule are then multiplied by weights attached to the connection of the input unit to the hidden unit. The minimum of those values is its final input. In each hidden unit a special monotonic membership function (see also Section 5) representing the conclusion of the rule is stored. Because of the monotonicity of this function the crisp output value belonging to the minimum membership value can be easily calculated by the inverse function. This value is multiplied with the weight of the connection from the hidden unit to the output unit. The output value is then calculated as a weighted average of all rule conclusions.

The AEN tries to predict the system behavior. It is a feed–forward neural network with one hidden layer, that receives the system state as its input and an error signal r from the physical system as additional information. The output $v[t, t']$ of the network is viewed as a *prediction of future reinforcement*, that depends of the weights of time t and the system state of time t', where t' may be t or $t + 1$. Better states are characterized by higher reinforcements.

The weight changes are determined by a reinforcement procedure that uses the outputs of the ASN and the AEN. The ARIC architecture was applied to cart–pole balancing and it was shown that the system is able to solve this task [1].

We call the Berenji's system *a fuzzy control oriented neural network.* It is a system that consists of more or less standard neural networks, where one network employs a fuzzy aspect in using membership functions within its units and a fuzzy inference scheme. This approach shows the following advantages: The system makes use of an existing rule base, it has not to learn from scratch. One main part of the system is build according to those fuzzy–if–then rules, and so its structure can be easily interpreted. Finally, the system is able to learn to improve its control actions by using an error signal from the controlled physical system.

From our point of view ARIC has a few disadvantages that we hope to overcome with our approach presented in the next section. The first point of our criticism is that the changes in the weights of the fuzzy inference part of the ASN are difficult to interpret. A change in the weight connecting an input unit and a hidden unit means a change in the respective membership function of the input unit. It is possible that the same input unit is connected to several hidden units, meaning there are several rules which share common linguistic values in their antecedents. The problem is that each connection has a *different* weight. This results in the fact, that two rules sharing a common fuzzy set may have different interpretations of the same variable state because of different membership functions. However, to retain the semantics of the fuzzy rule base, we want each rule to have the same interpretation, and a change in the membership functions during a learning process should be the same for all rules.

Finally we criticize the use of the so called "probability" used for the "stochastic action modification" and the concept of the AEN. It is not obvious why the same weights are on the one hand useful to refine the fuzzy rule base and can be used on the other hand to calculate a "probability" for changing the output of the fuzzy inference network. The semantics of this value and the modification are not clear. The evaluation of the controller performed by the AEN is a computation based on the system state and the system error that cannot be traced. This value is used as an error signal to change the weights, but has also unclear semantics.

ARIC is a working neural network architecture making use of fuzzy control aspects to its advantage but still has the disadvantage of neural systems that do not allow the semantical interpretation of certain steps in the information processing.

In [2] Berenji and Khedkar present an enhancement of this approach called GARIC (generalized ARIC). There they have overcome the disadvantage of different interpretation of linguistic values by refraining from weighted connections in the ASN. The fuzzy sets are modelled as nodes in the network, and learning is achieved by changing parameters within these nodes that determine the shape of the membership functions. By defining a differentiable *softmin* function and a special defuzzification scheme (*localized mean of maximum*) they can use a different learning algorithm that allows to use any form of membership functions in the conclusions. The stochastic modification scheme is still in use for this new approach.

5   Neural Network Oriented Fuzzy Control

In this section we present our approach to a combination of neural networks and fuzzy controllers (see also [5]). Our main concern is to keep the structure of the fuzzy controller that is determined by the fuzzy rules. We think of these rules as a piece of structural knowledge that gives us a roughly correct representation of the system to be controlled. We consider it a natural approach to define a fuzzy error for our system, which, according to its structure, we may call neural fuzzy controller. This enables us to define a reinforcement learning algorithm that is described at length in [12]. Training a fuzzy controller with such a learning procedure allows us to keep track of the changes and to interpret the modified rules.

We consider a dynamical system S that can be controlled by one variable C and whose state can be described by n variables $X_1, \ldots, X_n$, i.e. we have a multiple input – single output system. The variables are modelled by membership functions, and the control action that drives the system S to a desired state is described by the well-known concept of fuzzy if-then rules [21].

To overcome the problem of defuzzification, and to determine the individual part that each rule contributes to the final output value, we use Tsukamoto's monotonic membership functions, where the defuzzification is reduced to an application of the inverse function [1, 11]. Such a membership function μ is characterized by two points a, b with $\mu(a) = 0$ and $\mu(b) = 1$, and it is defined as

$$\mu(x) = \begin{cases} \frac{-x+a}{a-b} & \text{if } (x \in [a,b] \wedge a \leq b) \vee (x \in [b,a] \wedge a > b) \\ 0 & \text{otherwise} \end{cases}$$

The defuzzification is carried out by

$$x = \mu^{-1}(y) = -y(a-b) + a$$

with $y \in [0, 1]$.

For our purposes we only need to restrict ourselves to monotonic membership functions to represent the linguistic values of the output variable. For the input variables the usual triangular, trapezoidal etc. membership functions can be chosen.

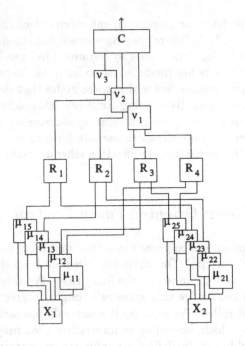

Fig. 1. The structure of the neural fuzzy controller

An example for the structure of our neural fuzzy controller is depicted in figure 1. The nodes X_1 and X_2 represent the input variables, and deliver their crisp values to their μ-modules which contain the respective membership functions. The μ-modules are connected to the R-modules which represent the fuzzy if-then rules. Each μ-module gives to all of its connected R-modules, the membership value $\mu_{ij}(x_i)$ of its input variable X_i. The R-modules use a t-norm (min-operation in this case) to calculate the conjunction of their inputs and pass this value forward to one of the ν-modules, which contain the membership functions representing the output variable. By passing through the ν-modules these values are changed to the conclusion of the respective rule. This means the implication (min-implication in this case) is carried out to obtain the value of the conclusion, and because monotonic membership functions are used, the ν-modules pass pairs $(r_l, \nu_k^{-1}(r_l))$ to the C-module, where the final output value is calculated by

$$
c = \frac{\sum\limits_{i=1}^{n} r_i \nu_{R_i}^{-1}(r_i)}{\sum\limits_{i=1}^{n} r_i},
$$

where n is the number of rules, and r_i is the degree to which rule R_i has fired.

As one can easily see, the system in figure 1 resembles a feedforward neural network. The X-, R-, and C-modules can be viewed as the neurons and the μ- and ν-units as the adaptable weights of the network. The fact that one μ-module can be connected to more than one R-module is equivalent to connections in a neural network that share a common weight. This is very important, because we want each linguistic value to be represented by only one membership function that is valid for all rules. By this restriction we retain the structural knowledge that we put into the system by defining the rules. In other neural fuzzy systems this fact is not recognized [1, 9] and it is possible that one linguistic value is represented by different membership functions.

Our goal is to tune the membership functions of the controller by a learning algorithm according to a measure that adequately describes the state of the plant under consideration. The desired optimal state of the plant can be described by a vector of state variable values. But usually we are content with the current state if the variables have roughly taken these values. And so it is natural to define the goodness of the current state by a membership function from which we can derive a fuzzy error [12] that characterizes the performance of our neural fuzzy controller.

Consider a system with n state variables $X_1, \ldots, X_n$. We define the fuzzy-goodness G_1 as

$$G_1 = \min\left\{\mu_{\text{opt}}^{(1)}(x_1), \ldots, \mu_{\text{opt}}^{(n)}(x_n)\right\},$$

where the membership functions $\mu_{\text{opt}}^{(i)}$ have to be defined according to the requirements of the plant under consideration.

In addition of a near optimal state we also consider states as good, where the incorrect values of the state variables compensate each other in a way, that the plant is driven towards its optimal state. We define the fuzzy-goodness G_2 as

$$G_2 = \min\left\{\mu_{\text{comp}}^{(1)}(x_1, \ldots, x_n), \ldots, \mu_{\text{comp}}^{(k)}(x_1, \ldots, x_n)\right\}$$

where the membership functions $\mu_{\text{comp}}^{(j)}$ again have to be defined according to the requirements of the plant. There may be more than one $\mu_{\text{comp}}^{(j)}$ and they may depend on two or more of the state variables.

The overall fuzzy-goodness is defined as

$$G = g(G_1, G_2),$$

where the operation g has to be specified according to the actual application [12]. The fuzzy-error that is made by our neural fuzzy controller is defined as

$$E = 1 - G.$$

The learning algorithm depending on this fuzzy error works for each fuzzy rule in parallel. Each rule R_i knows the value r_i of the conjunction of its antecedents and the (crisp) value c_i of its conclusion. After the control action has been determined by the controller and the new state of the plant is known, we propagate the fuzzy error E and the current values of the state variables to each

R-module. If the rule has contributed to the control output, i.e. $r_i \neq 0$, it has to evaluate its own conclusion. According to the current state of the plant the rule can decide, whether its conclusion would drive the system to a better or to a worse state. For the first case the rule has to be made more sensitive and has to produce a conclusion that increases the current control action, i.e. makes it more positive or negative respectively. For the second case the opposite action has to be taken.

Consider that we are using Tsukamoto's monotonic membership functions. Each membership function can be characterized by a pair (a,b) such that $\mu(a) = 0$ and $\mu(b) = 1$ hold. A rule is made more sensitive by increasing the difference between these two values in each of its antecedents. That is done by keeping the value of b and changing a. That means the membership functions are keeping their positions determined by their b-values, and their ranges determined by $|a - b|$ are made wider. To make a rule less sensitive the ranges have to be made smaller. In addition to the changes in its antecedents, each firing rule has to change the membership function of its conclusion. If a rule has produced a good control value, this value is made better by decreasing the difference $|a - b|$, and a bad control value is made less worse by increasing $|a - b|$.

The rules change the membership functions by propagating their own rule-error

$$e_{R_i} = \begin{cases} -r_i \cdot E & \text{if } \text{sgn}(c_i) = \text{sgn}(c_{\text{opt}}) \\ r_i \cdot E & \text{if } \text{sgn}(c_i) \neq \text{sgn}(c_{\text{opt}}) \end{cases}$$

to the connected μ- and ν-modules. The changes in the membership functions of the conclusions (ν-modules) are calculated according to

$$a_k^{\text{new}} = \begin{cases} a_k - \sigma \cdot e_{R_i} \cdot |a_k - b_k| & \text{if } (a_k < b_k) \\ a_k + \sigma \cdot e_{R_i} \cdot |a_k - b_k| & \text{otherwise,} \end{cases}$$

where σ is a learning factor and R-module R_i is connected through ν_k to the C-module. If ν_k is shared, it is changed by as much R-modules as are connected to the C-module through this membership function. For the membership functions of the antecedents (μ-modules) the following calculation is carried out:

$$a_{jk_j}^{\text{new}} = \begin{cases} a_{jk_j} + \sigma \cdot e_{R_i} \cdot |a_{jk_j} - b_{jk_j}| & \text{if } (a_{jk_j} < b_{jk_j}) \\ a_{jk_j} - \sigma \cdot e_{R_i} \cdot |a_{jk_j} - b_{jk_j}| & \text{otherwise,} \end{cases}$$

where the X-module X_j is connected to the R-module R_i through the membership function μ_{jk_j}, with $k_j \in \{1, \ldots, s_j\}$, and s_j is the number of linguistic values of X_j. If μ_{jk_j} is shared, it is changed by as much R-modules as X_j is connected to through this μ-module.

Our fuzzy error propagation learning algorithm is a reinforcement learning algorithm where each R-module learns according to the extent it has fired under the given circumstances and according to an error describing the performance of the control system. The neural fuzzy controller has not to learn from scratch, but knowledge in the form of fuzzy if-then rules is coded into the system. The learning procedure does not change this structural knowledge. It tunes the membership functions in an obvious way, and the semantics of the rules is not blurred by any semantically suspicious factors or weights attached to the rules.

The algorithm finds the most appropriate membership functions for the current control situation, while the meaning of each rule is not changed. If the learning is successful, this means that the rule base is suitable for the control task, and the changes in the fuzzy sets are due to an unsuitable modeling of the linguistic values by the a priori membership functions. It is not possible that two rules use different fuzzy sets describing the same linguistic value. That means we can interpret the new membership functions to give us the best fuzzy sets for the current task.

The neural fuzzy controller has been tested in a simulation using a simplified version of the inverted pendulum described by the differential equation

$$(m + \sin^2 \theta)\ddot{\theta} + \frac{1}{2}\dot{\theta}^2 \sin(2\theta) - (m + 1)\sin \theta = -F \cos \theta.$$

The controller was already able to balance the pendulum using the initial rule base and membership functions without learning. But it was not able to maintain an angle of roughly $\theta = 0$, and it was not able to cope with extreme starting positions (e.g. $\theta = 20°$ and $\dot{\theta} = 60°$ per second). After the learning process has been switched on the performance of the controller increased. It was able to maintain an angle of approximately zero and to cope with starting positions of e.g. $\theta = 30°$ and $\dot{\theta} = 60°$ per second [12].

An extension of this model is presented in [15]. There the system starts to learn from scratch, i.e. it determines the fuzzy rule base while it learns. The idea is to initialize the network with all rules that can be constructed out of all combinations of input and output membership functions. During the learning process all hidden nodes (rule nodes) that are not used or produce counterproductive results are removed from the network. We call this system fuzzy neural network because its initial state resembles more a neural network than a fuzzy controller. But after the learning process the system can of course still be interpreted in terms of a fuzzy controller.

The learning procedure is divided in three phases:

1. During phase I all rule nodes that have produced a counterproductive result, i.e. a negative value where a positive value is required and vice versa, are deleted instantly. Furthermore each rule node maintains a counter that is decremented each time the rule node does not fire, i.e. $r_j = 0$. In the other case, i.e. $r_j > 0$, the counter is reset to a maximum value.

2. At the beginning of phase II there are no rule nodes left that have identical antecedents and consequences that produce output values of different directions. To obtain a sound rule base, from each group of rules with identical antecedents only one rule node must remain. During this phase the counters are evaluated and each time a counter reaches 0 the respective rule node is deleted. Furthermore each rule node now maintains an individual error value that accumulates a fuzzy transition error

$$E_t = 1 - min\left\{\tau_i(\Delta x_i)|i \in \{1, \ldots, n\}\right\},$$

where Δx_i is the change in variable X_i, and τ_i is a membership function giving a fuzzy representation of the desired change in the respective variable. If the rule produced a counterproductive result, E_t is added unscaled, and if not, E_t is weighted by the normalized difference between the rule output value and the control output value of the whole network. At the end of this phase from each group of rule nodes with identical antecedents, only the node with the least error value remains, all other rule nodes are deleted. This leaves the network with a minimal set of rules needed for the control task.

3. During phase III the performance of the fuzzy neural network is enhanced by tuning the membership functions as it is described above.

At the end of the learning process the remaining rule nodes identify the fuzzy if-then-rules that are necessary to control the dynamical system under consideration, and the fuzzy weights represent the membership functions that suitably describe the linguistic values of the input and output variables for this situation. A simulation of this model applied to the inverted pendulum gave first promising results. The system reduced the number of rules from 512 to a number between 20 and 40 depending on the internal parameters. In most of the simulation runs the resulting network was able to balance the pendulum. To present final results still more tests have to be carried out.

6 Conclusions

We have discussed four possible combinations of neural networks and fuzzy controllers. The first two approaches use stand–alone neural networks to tune membership functions or to create fuzzy–if–then rules. The remaining concepts both try to build hybrid systems that benefit from the advantages of both worlds. Berenji's ARIC architecture is a special neural network system that still has some semantical problems. We tried to overcome these problems with our neural fuzzy controller that uses a learning algorithm based on a fuzzy error measure. The structure of the controller resembles a neural network and the fuzzy error propagation is a reinforcement learning procedure as used for certain kinds of connectionistic systems. Simulations of the controller used to balance an inverted pendulum have shown that the learning procedure improves the behavior of the fuzzy controller and is able to handle extreme situations where the non-learning controller fails [12, 13]. The learning algorithm starts from a predefined rule base, and it does not change the structural knowledge encoded in these rules. It leaves the semantics of each rule intended by the user unchanged, but removes the errors caused by an inaccurate modelling by changing the fuzzy sets. The results of the learning procedure can be interpreted easily.

References

1. H.R. Berenji: A Reinforcement Learning-Based Architecture for Fuzzy Logic Control. Int. J. Approximate Reasoning **6** (1992), 267-292.

2. H.R. Berenji, P. Khedkar: Learning and Tuning Fuzzy Logic Controllers Through Reinforcements. IEEE Trans. Neural Networks 3 (1992), 724-740.
3. J.C. Bezdek, S.K. Pal (eds.): Fuzzy Models for Pattern Recognition. IEEE Press, New York (1992).
4. J.C. Bezdek, E.C. Tsao, N.K. Pal: Fuzzy Kohonen Clustering Networks. Proc. IEEE Int. Conf. on Fuzzy Systems 1992, San Diego (1992), 1035-1043.
5. P. Eklund, F. Klawonn, D. Nauck: Distributing Errors in Neural Fuzzy Control. Proc. 2nd Int. Conf. on Fuzzy Logic and Neural Networks, IIZUKA'92, Iizuka (1992), 1139-1142.
6. I. Hayashi, H. Nomura, H. Yamasaki, N. Wakami: Construction of Fuzzy Inference Rules by NFD and NDFL. Int. J. Approximate Reasoning 6 (1992), 241-266.
7. K. Hirota: Survey of Industrial Applications of Fuzzy Control in Japan. Proc. IJCAI-91 Workshop on Fuzzy Control, Sydney (1992), 18-20.
8. H. Ichihashi: Iterative Fuzzy Modelling and a Hierarchical Network. In: R. Lowen, M. Roubens (eds.): Proc. 4th IFSA Congress, Engineering, Brussels (1991), 49-52.
9. J.-S.R. Jang: Fuzzy Modelling Using Generalized Neural Networks and Kalman Filter Algorithm. Proc. 9th Nat. Conf. on Artificial Intelligence, AAAI-91, MIT-Press, Menlo Park (1991), 762-767.
10. B. Kosko: Neural Networks and Fuzzy Systems. Prentice-Hall, Englewood Cliffs (1992).
11. C.C. Lee: Fuzzy Logic in Control Systems: Fuzzy Logic Controller. IEEE Trans. Syst. Man Cybern. 20 (1990), Part I: 404-418, Part II: 419-435.
12. D. Nauck, R. Kruse: A Neural Fuzzy Controller Learning by Fuzzy Error Propagation. Proc. NAFIPS'92, Puerto Vallarta, (1992), 388-397.
13. D. Nauck, R. Kruse: Interpreting Changes in the Fuzzy Sets of a Self-Adaptive Neural Fuzzy Controller. Proc. 2nd Int. Workshop in Industrial Fuzzy Control and Intelligent Systems IFIS'92, College Station (1992) 146-152.
14. D. Nauck, F. Klawonn, R. Kruse: Fuzzy Sets, Fuzzy Controllers, and Neural Networks. Wissenschaftliche Zeitschrift der Humboldt-Universität zu Berlin, R. Medizin 41 (4) (1992), 99-120.
15. D. Nauck, R. Kruse: A Fuzzy Neural Network Learning Fuzzy Control Rules and Membership Functions by Fuzzy Error Backpropagation. Proc. IEEE Int. Conf. on Neural Networks ICNN'93, San Francisco (1993).
16. H. Nomura, I. Hayashi, N. Wakami: A Learning Method of Fuzzy Inference Rules by Descent Method. Proc. IEEE Int. Conf. on Fuzzy Systems 1992, San Diego (1992), 203-210.
17. W. Pedrycz, H.C. Card: Linguistic Interpretation of Self-Organizing Maps. Proc. IEEE Int. Conf. on Fuzzy Systems 1992, San Diego (1992), 371-378.
18. S. Shao: Fuzzy Self-Organizing Controller and its Application for Dynamic Processes. Fuzzy Sets and Systems 26 (1988), 151-164.
19. H. Takagi, I. Hayashi: NN-Driven Fuzzy Reasoning. Int. J. Approximate Reasoning 5 (1991), 191-212.
20. H. Takagi, N. Suzuki, T. Koda, Y. Kojima: Neural Networks Designed on Approximate Reasoning Architecture and their Applications. IEEE Trans. Neural Networks 3 (1992), 752-760.
21. L.A. Zadeh: Outline of a New Approach to the Analysis of Complex Systems and Decision Processes. IEEE Trans. Syst. Man Cybern. 3 (1973), 28-44.

A reinforcement learning algorithm based on 'safety'

A. Nowé, University of Brussels,
associate student at Queen Mary and Westfield College, University London
R. Vepa, Queen Mary and Westfield College, University London

Abstract In this paper, a reinforcement learning algorithm is presented which is used to implement a fuzzy controller model for a given control problem based on the concept of 'safety'. A concept of 'safety' is postulated and learned iteratively. It embodies the notion of cost as well as performance. The fuzzy controller model which is based on this notion of safety is closely related to the controllability of the system and takes the physical constraints of the controller into account.

1 Introduction

Fuzzy control systems have already proven to be very useful. In recent years, they have been applied to a broad variety of systems. Fuzzy controllers are well suited for controlling non-linear and/or time variant systems. Moreover, they are reliable and robust [1].

For a wide range of applications, human experts are available who can provide their control expertise in linguistic rules and a straightforward implementation of the control strategy is possible. However, finding the right parameters of the fuzzy rules, often turns out to be very difficult. For other applications, no linguistic rules are at hand. In both cases, the automatic acquisition of fuzzy rules is an important problem.

In the literature two major categories of solutions to the acquisition problem can be distinguished. The first category assumes that somehow a control function is known. The control function can, for example, be described by a set of *state/action*-pairs. The acquisition problem is here reduced to an *approximation* problem. I.e. finding the right parameters for the fuzzy sets constituting the fuzzy rules, such that the error between the control function yield by the fuzzy controller and the given control function is minimised.

Approaches belonging to this first category are often Neural Net approaches. Examples of the application of gradient descent/back propagation techniques are given by [2,3,4,5]. An Genetic Algorithm approach is this category can be found in [6].

The second category also supposes that the notion of 'good control' is given, however, not as a set of state-action pairs, but as the result of applying an appropriate control strategy. A typical strategy is based on minimising a certain cost functional. Of course this does not always has to result in good performance. In classical optimal control choosing a quadratic cost functional [7,8] usually results in a stable controller with good performance properties. Thus the choice of a cost index is important and the whole purpose of a learning controller is to identify the best cost index that results in a good performance. Thus if one has a good performance indicator it is possible in principle to update the cost index to achieve good performance.

This distinction between performance and cost has not been recognised by [7,8] . Barto [9,10] was one of the first to recognise the need for an assessment of performance of the controller independent of the control strategy and this they do by using a reinforcement learning strategy. Recently Berenji and Khedkar [11,12] have presented a fuzzy learning controller based on the independent evaluation of performance of the control action and the selection strategy. Kosko and Kong [13,14] introduce the idea of differential competitive Hebbian learning based on classical Hebbian learning concepts as a strategy for synthesising controllers. However it is important in their approach to have a notion of what is a 'good' control action when the desired performance is achieved. In the context of fuzzy rule based systems one must have a rule in the rule base of the form IF error = 0 THEN control U is U $_0$. Thus the only performance criterion used is the requirement that the error must be zero.

It is quite apparent form these previous approaches that there is a need to assess both the performance and the cost.

2 A new Approach

An alternate approach to learning control is presented in this paper. The methodology proposed not only does not require an a priori knowledge of a satisfying *state/action*-pairs but also does not require an a priori knowledge of a 'good' performance index. On the contrary, the learning algorithm is based on the following idea.

"If we want to build a robust controller which keeps a certain process within a predefined viability domain, it is important to design the controller such that it brings and keeps the system into a safe state."

A *safe state* is a state where the system is not very sensitive for disturbances coming from the environment and consequently does not easily goes out of control. Our learning algorithm locates the safe regions and tunes the controller such that it brings and keeps the system into the safest region.

This is the basic control strategy involved. A quantitative measure of safety is developed and the controller drives the system so as to maximise this measure. Inherent in this concept is the fact that in doing so it is forcing the system to an equilibrium point, that is, a point in the state space where the measure of safety is a maximum. Thus the stability of the equilibrium of the system is ensured provided the quantitative measure of safety is a maximum at a certain point of interest in the state space. Further the quantitative measure of safety is chosen so as to reflect the need for a wide spectrum of control actions being available to the controller at the equilibrium point. Thus the quantitative measure of safety embodies both the concepts of cost as well as performance and is learned iteratively.

The viability domain determines the minimal demands to which the system indefinitely has to obey. The system is considered to be out of control when it leaves the viability domain. J.P. Aubin has developed a complete mathematical framework for non-linear control problems with state constraints [15,16]. As stated in [17] the constraint of viability might relax the objective to follow a referential trajectory for processes which are either too complex or interacting with an open and formally hard and unpredictable environment. On the other hand this constraint fits the objective of reliability or safety which is generally specified via the establishment of a bounded zone for some of the variables of the process. For the time being the work of J.P. Aubin is strictly theoretical and can not yet be applied in the engineering sense.

The rest of this paper is organised as follows. In Section 3, the concept of *fuzzy controller model* and the concept *safety* are introduced. The *measure of safety* is defined in the following section. In Section 5, the algorithm which learns a fuzzy controller model is presented. In Section 6, the learning algorithm is evaluated. An application is described in Section 7. Finally the conclusions are given in Section 8.

3 The fuzzy controller model and 'safety'

It has been stated by J.P. Aubin that the spectrum of viable control actions in a state, is an indicator of the "robustness" of the system. A control action is said to be viable if the system remains within the viability domain subsequent to the application of the control action. The wider the spectrum is, the better the system handles unexpected disturbances coming from the environment [15,16]. So, it is important to search for safe regions within the viability domain, and to design the controller such that the system is driven into a safe region.

To model the spectrum of viable control actions, fuzzy sets may be used. By using fuzzy sets, we can not only discriminate the viable control actions from the non-viable control actions, but also assess that one control action is better than another control action.

The model of a fuzzy controller is defined as a fuzzy function f which maps a fuzzy region, modelled by a fuzzy set over the state space, I_j, to a spectrum of control actions modelled by a fuzzy set over the possible control actions, O_j. The image $f\left(I_j\right)$ of each fuzzy region I_j expresses a fuzzy restriction O_j on the possible control actions for that region.

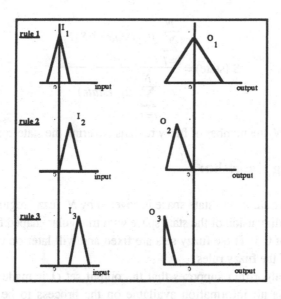

Figure 1 : The controller model

Figure 1 depicts an example of a fuzzy controller model. The model constraints 3 fuzzy rules. Both the input space and output space are considered to be one-dimensional.

Based on the concept of 'safety' presented earlier, the fuzzy region I_1 can be said to be more safe than the region I_3.

The basis of a fuzzy controller is a fuzzy rule base, consisting of several rules of the form IF A_1 AND ... AND A_n THEN B. The fuzzy regions which are described using fuzzy sets correspond to the antecedents of the fuzzy rules. The consequents of the rules are modelled by the images of the corresponding antecedents.

4 The measure of 'safety'

The safety of a fuzzy region is reflected in the spectrum of the corresponding image. The wider the spectrum, the safer the region. It is therefore proposed to measure the safety of a fuzzy region I_j by taking the integral of the membership function μ_{oi} of the corresponding image O_i .

$$S(I_j) = \int_O \mu_{oi} \text{ , with } O \text{ the set of control actions.}$$

During the learning we also need to measure the safety of a crisp state. Since a crisp state generally belongs to several fuzzy regions, the safety measure is defined as the weighed average of the safety measures of all the fuzzy regions to which the crisp state belongs.

$$S\ (state) = \frac{\sum\limits_{i=1}^{N} \mu_{I_i}\ (state) * S\ (I_j)}{\sum\limits_{i=1}^{N} \mu_{I_i}\ (state)}$$

with N the number of fuzzy regions covering the state space.

5 The learning algorithm

Before the learning starts, the state space is covered by N fuzzy regions. This is done by covering each dimension of the state space with triangular shaped fuzzy sets which cross at a grade of 0.5. These fuzzy sets are fixed and will later on be considered as the antecedents of the fuzzy rules.

The learning algorithm supposes that the output set O is made discrete. In the beginning there is no information available on the process to be controlled and consequently each control action is equally possible, e.g. a possibility 1/2 is therefore assigned to each control action. This means that the consequent O_j, $1 \leq j \leq N$, $\forall o \in O$, $\mu_{O_j}\ (o) = 1/2$.

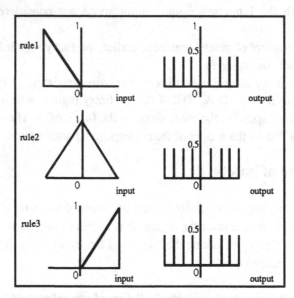

Figure 2 : An initial state

Figure 2 shows a possible initial state for a SISO system. The state space is covered by 3 fuzzy regions I_1 , I_2 en I_3 . Thus, the fuzzy controller will eventually contain 3 rules. The set of control actions is discretised into 9 equidistant values.

Initially, each membership grade of every output o to O_1, O_2 and O_3 is equal to 1/2.

The task of the learning algorithm is to adjust the membership functions of the conclusions O_i of the N fuzzy rules such that :

- A control action o which brings a fuzzy region I_j out of control needs to be excluded. This results in a 0 membership grade in the corresponding O_j.
- The best control action o for a fuzzy region I_j is the action which brings the system into the safest region that can be reached from I_j. It is rewarded with a membership grade equal to 1 in the corresponding image O_j.
- The membership grade of other control actions is between 0 and 1, related to the safety of the state the system is brought into. The larger the safety, the larger the membership grade.

To learn the membership functions μ_{oi} defining the consequents O_j, the following reinforcement learning algorithm is used.

Repeat until convergence is reached

- Generate a 'random' initial state i_t within the viability domain,

- Repeat

     - select a control action o_s, at random

     - apply the control action during a time period Δt to the system,

     - observe the new state i_{t+1}

     - change the membership grade of o_s of all consequents O_j with $\Delta O_j (o_s)$

$$\Delta O_j (o_s) = \mu_{I_j} (i_t) \times \alpha \times (R(i, o_s) - O_j (o_s)) \qquad (1)$$

     and

$$R(i, o_s) = \begin{cases} 0 & \text{if out of control} \\ \min\left\{ 1, \dfrac{SM(i_{t+1})}{maximal\ expectation\ for\ I_j} \right\} & \text{else} \end{cases}$$

- until out of control

In equation (1), $\alpha \in [0, 1]$ is the learning rate, when the learning is completed, then α may be set to zero, as no further modification of the rule base is needed.

The factor $\mu_{I_j}(i_t)$ reflects the fact that the stronger the i_t belongs to I_j the more representative the i_t -o_s relation is of for the fuzzy relation between I_j and O_j .The maximal expectation of I_j is defined as follows:

The control action which brings the system in the fuzzy region I_t into the safest fuzzy region I_{t+1} , that can be reached from I_t with the given range of control actions, needs to have the largest membership grade, i.e. 1. During the learning process, the membership functions of the O_j are constantly changed. This means that the safety $S(i_{t+1})$ may change at every time step. Hence a dynamical notion of the safest reachable regions is needed.

The solution is as follows: With each fuzzy region I_j a table of the safety measures of the 'n' of most recent states reached from a state i_j for which $\mu_{I_j}(i_j) > 0.5$ is computed. The maximal expectation of I_j is defined as the maximum of this table. (The number n is chosen to be twice the number of elements of the discretised output set O)

The learning strategy proposed is quite independent of the fuzzy inference methods that are used in fuzzy control. The combination of the antecedents could be done using any t-norm, such as the AND operator or the product operator. The results of a sample candidates indicate that the learning algorithm is quite insensitive to the choice of the t-norm.

In the following section, the results are shown for two different combinations of t-norms and t-conorms and compared with the performance of a Sugeno type controller, also based on the rulles obtained by the learned algorithm.

6 Evaluation

To illustrate the performance of the different fuzzy controllers, the well-known inverted pendulum problem is used . A cart is free to move along a one-dimensional track while a pole is free to rotate in the vertical plane of the cart and track. A force F, ranging from -12 Newton to +12 Newton is applied at discrete time intervals to the center of mass of the cart. The state of the inverted pendulum is described by two real valued variables θ and $\dot{\theta}$, respectively the angle of the pole with the vertical and the angular velocity of the pole. The system is out of control if the angle $\theta \notin$ [-12 degrees , +12 degrees].

In this experiment, the θ direction of the input space is covered by 5 fuzzy regions: NM_θ , NS_θ , ZR_θ , PS_θ and PM_θ. The $\dot{\theta}$ direction with 5 fuzzy regions: $NM_{\dot{\theta}}$, $NS_{\dot{\theta}}$, $ZR_{\dot{\theta}}$, $PS_{\dot{\theta}}$ en $PM_{\dot{\theta}}$. The set of available control actions is descretised in steps of 2 Newton. A learning rate $\alpha = 0.5$ and 10000 iteration steps were used.

The inverted pendulum is simulated on a computer using a time step of 0.1 seconds.

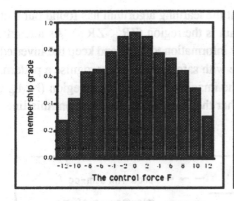

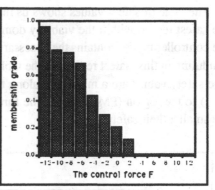

Figure 3 a : $O_{ZR_\theta, ZR_{\dot\theta}}$ Figure 3 b : $O_{ZR_\theta, ZR_{\dot\theta}}$

The fuzzy controller model contains 5x5 rules. As an example two of the consequents of the learned rules are shown in figure 3. In figure 3 a depicts the consequent $O_{ZR_\theta, ZR_{\dot\theta}}$. All control actions are viable since they all have a membership grade larger than zero. However the best control action is the action "O Newton". This action brings the region into safest reachable region, i.e. the region (ZR_θ, $ZR_{\dot\theta}$) itself.

As a second example the consequent $O_{NS_\theta, NS_{\dot\theta}}$ is shown. Controlling the (NS_θ, $NS_{\dot\theta}$) region is more difficult, some of the control action are not viable, their membership grade is zero. The best control action for this region is "-12 Newton".

	NM_θ	NS_θ	ZR_θ	PS_θ	PM_θ
$PM_{\dot\theta}$	5.8	4.3	2.8	0.8	0.0
$PS_{\dot\theta}$	8.3	7.8	6.5	4.1	1.5
$ZR_{\dot\theta}$	5.6	8.0	8.4	8.1	5.6
$NS_{\dot\theta}$	1.0	4.4	6.9	8.0	8.3
$NM_{\dot\theta}$	0.0	1.3	3.1	4.7	5.5

Table 1: The safety values

Table 1 shows the safety values for the 25 fuzzy regions. The values are obtained by summing the membership grades of all control actions in the corresponding output sets. The regions (PM_θ, $PM_{\dot\theta}$) and (NM_θ, $NM_{\dot\theta}$) are very unsafe. Their safety value is zero, which means that they are not controllable with the available range of control actions. Therefore these regions have to be avoided if the system has to be kept viable. On the contrary, as could be expected, the region (ZR_θ, $ZR_{\dot\theta}$) is a very safe region, it has the largest safety value. A large spectrum of viable control actions is related to this region.

The table of safety values shows us that the learning algorithm has found out that the safest region within the viability domain is the region (ZR $_\theta$ ZR $_{\dot\theta}$). As a result, the controller model contains the necessary information to drive and keep the inverted pendulum in this safest region. In the table with safety values we recognise a pattern. The safety values are a maximum along the imaginary line form the region (NM $_\theta$, PS $_{\dot\theta}$) to the region (PM$_\theta$, NS $_{\dot\theta}$). The further the regions are situated from this line, the smaller their safety value.

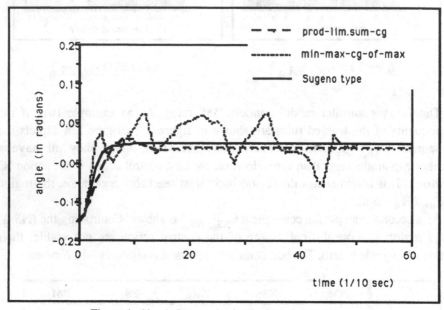

Figure 4 : Simulation results for the inverted pendulum

Figure 4 shows the performance of the 3 different fuzzy controllers obtained by taking different t-norms, t-conorms and defuzzification methods.

As we can see from the figure, the min-max-combination-rule with the centre of gravity of the maxima (min-max-cg-of-max) shows a moderate performance. After some oscillation the system settles in a very safe equilibrium i.e. the vertical position, which indeed belongs to the safest region (ZR$_\theta$, Z R $_{\dot\theta}$). The fuzzy controller obtained by taking the product as t-norm, the limited sum as t-conorm and centre of gravity as defuzzification procedure (prod-lim.sum-cg) shows a very good performance the safe equilibrium point is reached within a small rise time and almost no overshoot is observed. The best performance is yield by the Sugeno typed controller. This fuzzy controller contains rules of the following form:

IF $(\theta, \dot\theta)$ is I_j THEN o is o_i

with $o_i \in O$ such that $\mu_{o_i}(o_i) \geq \mu_{o_i}(o_j) \ \forall o_j \in O$

This means that only the best control action for each of the fuzzy conclusions is retained.

Whether these observations on the choice of t-norm, t-conorm and defuzzification method, may be generalised, is currently examined by the authors. We remark that most of the conclusion O_j obtained by the learning algorithm are asymetrical shaped fuzzy sets. This implies that studies reported in literature on the best choice of t-norm, t-conorm and defuzzification method, carried out for symetrical shaped conclusions, are not applicable here.

7 Application to Robot Manipulators

In the previous described experiment it is quite apparent that the safest position for the inverted pendulum is the vertical position. Notice however, that this information has not been provided to the learning algorithm. The learning algorithm autonomously determined the vertical position as a set point. This property of the learning algorithm makes it possible to utilise it for interesting applications for which it is not at all obvious what the desirable set point is. One of these applications is the resolution of redundancies which often occur for conventional robot manipulators. In case of redundancies it is not enough if one resolves the redundancy; rather it should be exploited fully to perform the task at hand dextrously and in a versatile manner. It is therefore important to evaluate relevant performance metrics to achieve this objective.

A number of techniques have been proposed in the literature for the resolution of the redundancy based on a variety of concepts. One of the more recent techniques proposed is based on configuration control where the redundancy of the manipulator was utilised to achieve an optimum redundant configuration directly in the task space, thereby also avoiding the inverse kinematic transformation from the task to the joint space. While the concept has been an extremely reasonable and realistic one, it is almost always essential that the optimal configuration must be specified in some form or the other by the user. Thus there is a need for an auxiliary or additional set point or task trajectory.

In most space-based robot systems, such as those envisaged for use in a space station, the concept is not entirely practical and it is essential that this additional task be set autonomously. Further the control techniques suggested in the literature for use with redundant manipulators have almost universally been model based.

The Fuzzy Learning algorithm proposed earlier in this paper is a candidate for a model free and autonomous approach to the problem. The autonomous configuration control scheme is based on the idea of Fuzzy compensation and is illustrated in Figure 5. A model free approach is also adopted for the primary task trajectory control.

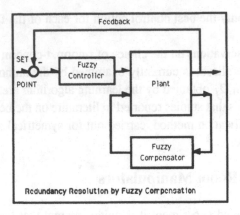

Figure 5 : The fuzzy Compensator Concept

The fuzzy compensation is based on the concept of 'safety' where the compensator seeks to function in such a way that the system's operating point in the state space has the widest possible spectrum of control actions available for implementation. Since some points in the multi-dimensional state space are constrained by the need for the task trajectory to be followed, the other points are so chosen such that a performance metric, which is a measure of the control actions, available is maximised. The fuzzy compensator thus ensures that the robot is operating at a point of maximum safety thus ensuring that it is in a stable configuration.

The concept has been applied by the authors to a dual link model of the UNIMATION PUMA 560 robot manipulator. This manipulator can be represented by an inverted double link configuration. Experiments in the joint space have already been carried out. It indicates that the fuzzy compensator, which controls the lower pole and is based on the learning algorithm, 'safely' compensates the task of the upper pole. Alternate configurations with the task trajectory defined in the Cartesian space are also being investigated at present.

8 Conclusion

In this paper, we have presented a reinforcement learning algorithm which learns a fuzzy controller model. The learning is based on the concept of safety. The quantitative measure of safety embodies both the concepts of cost as well as performance and is learned iteratively. The learning algorithm locates the safest regions within the viability domain and learns how to drive and keep the system at a safe equilibrium point. The equilibrium point, determined by the learning algorithm itself, is a point in the state space where the measure of safety is a maximum. By choosing an appropriate t-norm, t-conorm and defuzzification method a robust fuzzy controller is obtained.

It may be noted that the safety measure is a maximum along a line in the phase plane and the controller has the characteristics of a sliding mode controller. This relationship is discussed by the authors in [18].

References

[1]   Yamakawa T. "Stabilization on an inverted pendulum by high speed fuzzy logic controller hardware system", In *Fuzzy Sets and Systems*, 32: 161-180.1989.

[2]   Jang R "Rule extraction using genralised neural networks", In *Proceedings of the IFSA '91 Conference*, Brussels, 1991.

[3]   Yamaoka M. & Mukaidono M. "A learning method of fuzzy inference rules with a neural network" In *Proceedings of the IFSA '91 Conference*, Brussels, 1991.

[4]   Normura H., Hayashi I. & Wakami, N. "A self tuning method of fuzzy control by descent method", In *Proceedings of the IFSA '91 Conference*, Brussels, 1991.

[5]   Shen Z., Ding L. & Mukaidono M. "A self-learning approach for fuzzy simulating models", In *Proceedings of the International Conference on Fuzzy Logic and Neural Networks*,Iizuka, 1990.

[6]   Valenzuela-Rendon M. "The Fuzzy Classifier System: A Classifier System for Continuously Varying Variables", In *Proceedings of the Fourth International Conference on Genetic Algorithms*, Morgan Kaufmann, 1991.

[7]   Karr C.L. "Design of an Adaptive Fuzzy Logic Controller Using a Genetic Algorithm", In *Proceedings of the Fourth International Conference on Genetic Algorithms*, Morgan Kaufmann.

[8]   Bersini H. Génération Automatique de Systèmes de Commande Floue par les Métodes de Gradient et les Algorithmes Génétiques. Presented at the *DeuxièmesJournées Nationales sur les Apllications des Ensembles Flous*, Nîmes, 1992.

[9]   Barto A., Sutton R., Anderson C. "Neuronlike adaptive elements that can solve difficult learning control problems", *IEEE Transactions on Systems, Man and Sybernetics*, 13(5), 1983

[10] Barto A. Connectionist Learning for Control", In *Neural Networks for Control* (Thomas Miller III, Richard Sutton and Paul Werbos eds.) MIT Press.

[11] Berenji H.R. "A reinforcement Learning-Based Architecture for Fyzzy Logic Conrol", In *International Journal of Approximate Reasoning*, Vol 6,pp 267-292., 1992.

[12] Berenji H.R.and Khedkar P. "Learning and Tuning Logic Contollers Through Reinforcements, In *IEEE Transactions on Neural Networks*, Vol 3, No 5, 1992.

[13] Kosko B. & Kong S. "Adaptive Fuzzy Systems for Backing up a Truck-and-Trailer", In *IEEE transactions on Neural Networks*, Vol 3, No 2, March 1992.

[14] Kosko B. "Neural Networks and Fuzzy Systems : A Dynamical System Approach to Machine Intellingence", Englewood Cliffs, NJ : Prentice-Hall, 1992.

[15] Aubin J.P. "A survey of viability theory", In *SIAM J. Control and Optimization*, Vol. 28, No. 4, pp. 789-788, July 90.

[16] Aubin J.P. "Viability Theory", Systems &Control: Foundations & Applications, Brikhäuser, 1991.

[17] Bersini H. "Reinforcement Learning and Recruitement Mechanism for Adaptive Distributed Control", In *Proceedings of the 1992 Conference of AI in Real Time Control.*, 1992.

[18] Vepa R. and Nowé A. "A synergistic approach to learning fuzzy logic control rules. Self-organising fuzzy compensation of autonomous, redundant robot manipulators." In *proceedings of the 1st IFAC International Workshop on Intelligent Autonomous Vehicles*, Southampton, 1992.

GAITS : Fuzzy Sets-Based Algorithms for Computing Strategies Using Genetic Algorithms

Mohamed Quafafou*, Mohammed Nafia**

(*)   IRIN, Université de Sciences et des Techniques de Nantes, France.
(**)   Digital Equipment Corporation, Sophia-Antipolis, France.

Abstract. This paper introduces a new method combining Fuzzy Logic and Genetic Algorithms to allow Intelligent Tutoring Systems to be efficient. We consider supervised teaching where the teacher assigns a *initial profile* and a *finale profile* to each student before starting the teaching. The system computes an optimal strategy which represents a way of evolving the student's knowledge from the initial profile. So, the student's knowledge will change progressively to reach the final profile where the teaching objective is judged to be reached. The paper deals with an example presenting a simulation of the result's of a good student and a bad one.

1 Introduction

The architecture of Intelligent Tutoring System (ITS) is based on several interacting components. The following four components are usually used [8] :

- The expert module;
- The instructional module;
- The student model;
- Interfaces.

But these modules became complex, because research and development of an ITS to be used in real world is an extremely lenghty process which requires appropriate people and funding. Besides, when we develop an industrial educational system we have to cope with different kind of problems because phenomena are complex to formulate and generaly imperfectly known [6].

Symbolic tools was developed allowing to ITSs to be powerful and well organized educational systems [1, 2]. We propose an alternative numeric method, based on fuzzy logic and genetic algorithms which allows to ITSs to be efficient and to provide a maximum of competence in a minimum of time.

The main body of the paper comprises four sections. In the section 2 we discuss the interest of fuzzy logic for ITS and we propose a model based on fuzzy sets. In the first part of the section 3 we give a brief introduction to genetic algorithms (GAs). Then, we describe algorithms which allow to compute the *refrence strategy* and simulate student's feedback and answers. Finally, we deal with an example comparing the *reference strategy* to results obtained by both a good and a bad student.

2 Fuzzy logic & ITS

Recently, fuzzy logic was applied to ITS [7] in order to improve the speed in decision support over rule based inference engines and to have a more experienced control of learning sequence. Fuzzy logic was also used to improve a supervised customized teaching using pedagogical rules which are defined by the author [5].

In this paper, we assume that the tutor manages basic elements (*knowledge elements*). Each one defines the subject to be teached. Besides, we have a set of dialogues which can be proposed to the student considering his level. They have a normalized description characterized by two sets of worthed knowledge elements :

- *prerequisite knowledge elements (LP)* : a dialogue is candidate to be proposed to the student (admissible dialogue) if and only if its prerequisite elements are satisfied;
- *teached knowledge elements (LT)* : the dialogue execution allows to enrich the knowledge the student has.

For example, let us consider four knowledge elements e_1, e_2, e_3 and e_4 and the dialogue L_1 defined by $LP_1 = (0\ 0\ .1\ .3)$ and LT_1 $(.2\ 0\ .4\ .7)$. Note that e_1 and e_2 are not prerequisite elements for L_1 (their levels are equal to 0 in LP_1) when only e_2 is not teached by this dialogue (its level is equal to 0 in LT_1). The dialogue L_1 is considered an admissible dialogue iff the level of e_3 and e_4 in the student model are respectively greater then .1 and .3.

Let us consider supervised teaching where the teacher assigns the following profiles to each student before starting the teaching :

- *initial profile I* : is the student knowledge presupposed model at the time he starts his training and corresponds to the class of the student belongs to;
- *final profile F* : is a final state which must be reached by the student and where the teaching objective is judged reached.

Let us consider the precedent four knowledge elements and define the "teaching interval" by $I = (0\ .1\ 0\ .2)$ and $F = (.6\ .7\ .3\ .5)$. So, the teaching session allows to change progressively the student's profile from the initial profile to the final one. We define a strategy as a way of evolving the student's profile from I to F. In the following section we show how to compute the strategy taking account all useful knowledge.

We can view the LP_i, LT_i, I and F as fuzzy sets. They correspond to fit vectors and each fit value measures partial set membership or degrees of elementhood [4] For example, the fit value $m_I(e_2)$ equal to .1 indicates that the knowledge element e_2 belongs only slightly to the fuzzy set I. Besides, the fit value $m_F(e_4)$ equal to .5 indicates that the knowledge element e_4 belongs to F as much as it does not. The value 0 indicates the abscence of the ith knowledge element e_i in the fuzzy set.

3 Student's and expert's strategies

This section outlines the operation of a basic genetic algorithm (GA) and describes the GA-based method adopted in this study to compute the expert's strategy (reference strategy). For more details on GAs, see [3].

3.1 Genetic Algorithm

The basic idea of a genetic algorithm is to model a *chromosome* for a particular species and what happens to the chromosome during successive generations, when the underlying rule of nature is the survival of the fittest.

A simple GA is composed of three operators : *reproduction, crossover, mutation*. But, before using this operators and for each particular problem we have to define a *scheme for coding of the parameter space* which corresponds, in effect, to a descritisation of the parameter space. There must be a mapping between the points of the descretised space and a chromosome, which is just a finite length binary string in the basic algorithm. Each chromosome has an associated fitness value which is calculated by the fitness function.

At the start of the algorithm an initial population is generated randomly. Once the members of the population have been generated their fitness values are calculated via the predefined fitness function. Each string is a representation of a solution to the problem and its fitness value is a measure of the goodness of the solution that it represents. The initial population will be transformed by the genetic operators (reproduction, crossover, mutation) into populations with higher fitness values.

Using the reproduction operator, individual strings are copied according to their fitness function. The crossover operator chooses pairs of strings at random and produces new pairs. The *crossover probability* is the probability that the offspring will be different. If crossover is to take place then an integer j is selected at random to determine the crossover position. For example consider strings A and B and suppose that $j = 1$ then the two new individuals A', B' will be :

$$A = 1\ 0\ 0\ 1\ 1\ 1\ 0|0 \qquad A' = 1\ 0\ 0\ 1\ 1\ 1\ 0|1$$
$$B = 0\ 1\ 0\ 1\ 0\ 1\ 1|1 \qquad B' = 0\ 1\ 0\ 1\ 0\ 1\ 1|0$$

After crossover, the mutation operation is performed for each bit in each chromosome and if a mutation occurs then the particular bit just flip it's value. The basic genetic algorithm consists of repeating the above three operations of reproduction, crossover and mutation to repeatedly obtain new generations of individuals with, one hopes, higher levels of fitness.

3.2 The reference strategy

Before starting the teaching session, the tutor must be able to compute a strategy which allows to reach the final profil. This strategy is a way of changing progressively the student's profile from the initial profile to the final one. We call it the *reference strategy*, because it allows to judge student's results. In order to define this strategy, we compute at each teaching step T the *reference profile* η_T taking into account the following considerations :

- the student knowledge must evolve progressively;
- the teaching objective must be satisfied with the maximum degree at each step;
- all useful knowledge at the step T must be considered.

These considerations can be stated in numerical terms as follows :

- minimize $d(\eta_T, \text{Current profile})$;
- minimize $d(\eta_T, \text{Final profile})$;
- minimize the average of $d(\eta_T, L_i)$ for all admissible dialogues L_i at the step T.

So, the objective function to minimize is :

$$S_T(\eta_T) = d(\eta_T,C) + d(\eta_T,F) + \frac{1}{\sum\limits_{i=1}^{k} w_i} * \sum_{i=1}^{k} w_i * d(\eta_T,L_i) \qquad (1)$$

where

k : the number of admissible dialogues at the step T;
η_T : the reference profile at the step T;
C : the current profile;
F : the final profile;
L_i: the ith admissible dialogue at the step T;
w_i: the weighting coefficient for L_i.
$d(X,Y)$: the distance between the two fuzzy sets X, Y.

w_i express the importance of the dialogue L_i. For example, w_i is higher than w_j if the dialogue L_i is more important than L_j, that is, if LT_i is more near to the final profile F than LT_j ($d(LT_j, F) < d(LT_i, F)$).

The task of the optimization module is to determine the fuzzy set η_T by computing n values $m\eta_T(e_i)$, so as to minimise $S_T(\eta_T)$. The GA module contains the following procedures :

- *Initialization* : This procedure randomly generates the starting strings. Each one is of the form $x_1\ x_2\ ...\ x_m.$, where $x_i = m\eta_T(e_i)$ and are themselves binary strings;
- *Fitness* : This procedure computes the fitness of each solution string. Fitness scaling [6] is adopted;
- *Selection* : It implements the gene reproduction function;
- *Crossover* : It performs the gene crossover operation;
- *Mutation* : It carries out the gene mutation operation.

The computing *reference strategy* algorithm is *ReferenceStrategy* shown in the following. Input to ReferenceStrategy is a set of dialogues, the initial profile and the final profile. Ouput of ReferenceStrategy is the reference strategy Ω_{expert} composed by fuzzy sets $\eta_1, \eta_2, ..., \eta_f$. Besides, the satisfaction of the teaching objective degree δ_T at the step T are given as results.

So, before starting the teaching session, the tutor computes the reference strategy using the ReferenceStrategy procedure. Firstly, admissible dialogues are selected taking into account the current profile (I). Then the profile reference η_1 is computed minimizing the function S_1. Considering η_1 we compute the satisfaction of the teaching objective degree $\delta_1 = \text{Degree}(\eta_1 \supset F)$. Then we update the current profile and we start a teaching cycle by selecting a new admissible dialogue. The teaching objective is considered satisfied at the step p if $\delta_p = 1$. So, the reference strategy is defined by $\eta_1, \eta_2, ..., \eta_p$.

> **Procedure** ReferenceStrategy;
> **Input :**   • a initial profile I.
>   • a final profile F.
>   • a set of dialogues { L_i, i = 1 ,.., n }
> **Output :**   ♦ Ω : reference strategy.
>   ♦ δ_i : satisfaction of the teaching objective degree at the step i.
>   **begin**
>   CurrentProfile <- I;
> Ω <- $\varnothing$;   { $\varnothing$ is the empty set }
>   **While** the final profile is not reached **do**
>   Select admissible dialogues { => the ith Context Ci = L_i i = 1 ,.., k }
>   Minimize the function S_i using G.A. { => the ith Reference profile η_i }
> Ω <- $\Omega \otimes \eta_i$;   { Added η_i to Ω }
>   ComputeDegrees;   { compute performance ratio δ_i }
>   UpdateCurrentProfile;   { update the current profile considering η_i}
>   **end;**   { while }
>   Diagnosis;
> **end;** {ReferenceStrategy}

When the teaching session starts the tutor uses the StudentStrategy which allows to simulate the student feedback and answers considering his level. It has the same Input as the ReferenceStrategy procedure. Output of StudentStrategy is the adapted reference strategy to the student $W_{student}$ composed by fuzzy sets $\eta_1, \eta_2, ..., \eta_f$. Besides, the satisfaction of the teaching objective degree δ_T at the step T are given as result. Output of StudentStrategy are the reference strategy $\Omega_{student} = (\eta_1, \eta_2, ..., \eta_f)$, the student's strategy $\Theta = (\zeta_1, \zeta_1, ..., \zeta_1)$ and the student's performance ratios $\delta_T = \text{Degree}(\eta_T \supset F)$ and $\Delta_T = \|\eta_T - \zeta_T\|$ at the step T.

The tutor selects admissible dialogues considering the initial profile which is represented by the list of concepts the student supposed to control. The system must choose only one among these admissible dialogues to be proposed to the student. To do so, the system computes the reference profile η_1 which will be reached at this step considering all admissible dialogues, the current profile, the final profile and minimizing the objective function S_1. Then, we choose the nearest dialogue L_1 to η_1 which will be proposed to the student at the first step.

The tutor simulates the student interactions and at the end of the execution of the dialogue L_1 it updates the knowledge elements level of the teaching domain taking into account the student's answers. So, the new student's profile is ζ_1. Then the system change the student model considering ζ_1 and selects a new set of admissible dialogues. Next, a reference profile η_2 is computed and a new dialogue L_2 is proposed to the student. After execution of L_2 the student's profile equals to ζ_2. So, it will change progressively to reach the final profile, where the teaching objective is judged to be reached.

```
Procedure StudentStrategy;
Input :      • a initial profile I.
             • a final profile F.
             • a set of dialogues { Lᵢ, i = 1 ,.., n }
Output :     ♦ Ψ : dialogues proposed to the student.
             ♦ Ω : reference results.
             ♦ Θ : results after the execution of the dialogue Lᵢ.
             ♦ δᵢ, Δᵢ : performance ratios.
begin
     CurrentProfile <- I;
     Ψ <- ∅;                       { ∅ is the empty set }
     Ω <- ∅;
     Θ <- ∅;
     While the final profile is not reached do
          Select admissible dialogues  { => the ith Context Ci = Lᵢ, i = 1 ,.., k }
          Minimize the function Sᵢ using G.A. {=> the ith Reference profile ηᵢ }
          Choose the CurrentDialogue;   { => Lᵢ will be proposed to the student }
          StudentResults(Lᵢ, StudentLevel); { simulate student's results => ζᵢ }
          Ψ <- Ψ ⊗ Lᵢ;                 { Added Lᵢ to Ψ }
          Θ <- Θ ⊗ ζᵢ;                 { Added ζᵢ to Θ  }
          Ω <- Ω ⊗ ηᵢ;                 { Added ηᵢ to Ω }
          ComputeDegrees;              { compute performance ratios δᵢ and Δᵢ }
          UpdateCurrentProfile;        { update the current profile considering ζᵢ}
     end; { while }
     Diagnosis;
end; {Strategy}
```

Consequently, we have $\Theta_{student} = \zeta_1 \otimes \zeta_2 \otimes ... \otimes \zeta_q$ and $\Psi_{student} = L_1 \otimes L_2 \otimes ... \otimes L_q$. We have also as result $\Omega_{student} = \eta_1 \otimes \eta_2 \otimes ... \otimes \eta_q$ where the ith element η_i is a fuzzy set representing the reference profil which can be reached by the student at the ith step. On the other hand the teaching objective is judged to be reached when the final profile is included in ζ_q.

In order to evaluate the student's results at the ith step of the formation we compute the following degrees :

• reaching final profil degree : $\delta_i = Degree(\zeta_i \supset F)$;
• error degree : $\Delta_i = \| \eta_i - \zeta_i \|$.

4 Preliminary results

Let us consider an simple example with ten dialogues (table I) which have the same importance ($w_i = 1$) and the Euclidean distance between fuzzy sets :

$$d^2(X,Y) = \sum_{i=1}^{n} [m_X(e_i) - m_Y(e_i)]^2$$

Table I
Dialogues description

	Knowledge elements	Dialogues									
		L_1	L_2	L_3	L_4	L_5	L_6	L_7	L_8	L_9	L_{10}
Prerequisite:LP	e_1	0	0	0	.1	.3	.3	.1	.1	.2	.2
	e_2	0	0	0	.1	.1	0	.2	.2	.5	.4
Teached: LT	e_1	.2	.1	.3	.4	.2	.4	.3	.3	.5	.5
	e_2	.2	.3	.4	.2	.2	.2	.3	.4	.6	.5

We assign a "teaching interval" defining the following initial and final profile :

- the initale profile I = (0, 0);
- the finale profile F = (.8, .8).

The GA parameters employed are as follows :

- population size = 100
- Length of a chromozome = 8
- crossover probability = 1.
- mutation probability = 0.01
- maximum number of generations = 30

Firstly, we call the procedure ReferenceStrategy to compute the reference strategy : ReferenceStrategy. The results are shown in the table II. The final profile is reached in three steps.

TABLE II
Description of the reference strategy

Step i	Context C_i	Reference profil η_i	δ_i
1	1,2,3	(.2, .27)	.29
2	2,3,4,7,8	(.53, .6)	.71
3	3,4,5,6,7,8,9,10	(.8, .8)	1.

Then, we simulate the teaching session with a bad student (-) and a good one (+) : StudentStrategy('-') and StudentStrategy('+'). The results are given in the table III.

Remark : The context C_i is the set of admissible dialogues at the step i.

TABLE III
Simulated teaching steps for a good student (+) and a bad one (-)

step i	Student's level	Context C_i	Reference profil η_i	L_i	ζ_i	Δ_i	δ_i
1	+	1,2,3	(.2, .27)	1	(.15, .2)	.086	.22
	-	1,2,3	(.2, .27)	1	(.07, .07)	.238	.09
2	+	2,3,4	(.47, .53)	2	(.23, .53)	.253	.42
	-	2,3	(.33, .4)	2	(.11, .21)	.291	.20
3	+	3,4,7,8,10	(.6, .73)	7	(.44, .73)	.160	.73
	-	3,4,7,8	(.4, .53)	7	(.25, .3)	.275	.35
4	+	3,4,5,6,8,9,10	(.8, .8)	5	(.6, .91)	.228	.87
	-	3,4,8	(.6, .7)	4	(.4 .39)	.369	.50
5	+	3,4,6,8,9,10	(.8, .8)	4	(.99,1.)	.283	1.
	-	3,5,6,8	(.73, .73)	5	(.48, .48)	.354	.60
6	-	3,6,8,10	(.8, .8)	6	(.64, .57)	.280	.75
7	-	3,8,9,10	(.8, .8)	3	(.73, .7)	.122	.9
8	-	8,9,10	(.8, .8)	8	(.88, .8)	.080	1.

For comparison, Fig. 1. shows the reference strategy and the simulated results of a good student (+) and a bad one (-). The reference strategy has been adopted as the expert strategy which is not given by the author before starting the teaching session but computed by the tutor. It represents an average strategy which allows to judge student's results.

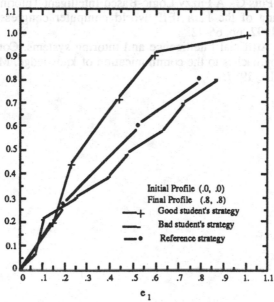

Fig. 1. Comparing profiles's evolution

5 Conclusion

This paper has described a new method for Fuzzy Intelligent Tutoring Systems. This method improves the "intelligence" of educational systems and their possibilities for assessment. Nowadays, we continue to study and to expand different aspects of the use of fuzzy logic in educational context. We have also to deal with the incomplete information problem. In fact, in practice it is complex to describe completely all pedagogical objects.

REFERENCES

1. Blandford A., An intelligent educational system to support the development of decision Making Skills Within engeneering design. In Proceedings of the International Conference on Computer Aided Training in Science and Technologie, Barcelona, 9-3 july 1990, pp. 183-190.
2. Elsom-cook M., Guided discovery tutoring, 1990, London: Chapman and Hill.
3. D.E. Goldberg, Genetic algorithms in search, optimization and machine learning, Addison-wesley, Reading, MA, 1989.
4. Kosko B., Neural networks and Fuzzy systems, Prentice Hall, Englewood Cliffs, NJ, 1992.
5. M. Quafafou, P. Prévot, CECIL II : An adaptive tutor based on fuzzy sets (submitted for publication), 1992.
6. M. Quafafou, P. Prevot, IEEE/ACM Conference on Developping and Managing Intelligent System Projects, Washington, D.C., March 29-31, 1993.
7. Regian W., Pitts G., A Fuzzy Logic-Based Intelligent Tutoring System (ITS). In Proceeding of the 12th IFIP world computer congress. Madrid, 7-11 September 1992, pp. 66-72.
8. Wenger E., Artificial Intelligence and tutoring systems. Computational and cognitive approaches to the communication of knowledge. Morgan Kaufman Publishers, Inc 1987.

Neural Networks and Genetic Algorithm Approaches to Auto-Design of Fuzzy Systems[o]

Hideyuki TAKAGI[1] and Michael LEE

Computer Science Division, University of California, Berkeley, CA 94720
takagi@cs.berkeley.edu, lee@cnmat.cnmat.berkeley.edu, FAX (510)642-5775

Abstract. This paper presents Neural Network and Genetic Algorithm approaches to fuzzy system design, which aims to shorten development time and increase system performance. An approach that uses neural network to represent multi-dimensional nonlinear membership functions and an approach to tune membership function parameters are given. A genetic algorithm approach that integrates and automates three fuzzy system design stages is also proposed.

1 Introduction

Fuzzy systems are frequently designed by hand. This poses two problems: (a) because hand design is time consuming, development costs can be very high; (b) there is no guarantee of obtaining an optimal solution. To shorten the development time and increase performance of fuzzy systems, there are two separate approaches: develop support tools and automatic design methods. The former includes developing environments to assist in fuzzy system design. Many environments are already commercially available. The latter approach involves introducing techniques to automate the design process. Though automatic design does not guarantee delivery of optimal solutions, they are preferable to manual techniques, because design is guided towards and an optimal solution by certain criteria.

There are three major design decisions to make when designing fuzzy systems:
(1) deciding the number of fuzzy rules,
(2) deciding the shape of the membership functions,
(3) deciding the consequent parameters.
Furthermore, two other decisions must be made:
(4) deciding the number of input variables,
(5) deciding the reasoning method.

(1) and (2) correspond to deciding how to cover the input space. They are highly dependent on each other. (3) corresponds to determining the coefficients of the linear equation in the case of the TSK (Takagi-Sugeno-Kang) model [1],

[o] This research is supported in part by NASA Grant NCC-2-275, MICRO State Program Award No.90-191, and EPRI Agreement RP8010-34. We would like to thank Prof. David Wessel and the Center for New Music and Audio Technologies at UC Berkeley for use of computing resources.
[1] The author is a Visiting Industrial Fellow at UC Berkeley and a Senior Researcher of Central Research Laboratories, Matsushita Electric Industrial Co., Ltd.

or determining consequent part membership functions in the case of the Mamdani model [2]. (4) corresponds to deciding the minimum set of relevant input variables needed to compute target decisions or control values. Techniques, such as backward elimination [4] or information criteria are often used to do in this task. (5) corresponds to deciding which fuzzy operator and defuzzification method to use. Although several operators and fuzzy reasoning methods have been proposed, there is no criteria for selecting them. [5] shows that dynamically changing the reasoning method according to the reasoning environment results in higher performance and fault-tolerance than any one fixed reasoning method.

Neural networks (more generally, gradient based models) and genetic Neural networks (most commonly gradient based) and genetic algorithms are used for auto-design of fuzzy systems. The neural network based methods are mainly used to design membership functions. There are two major methods;

(a) Direct Multi-Dimensional Membership Functions Design:
This method first decides the number of rules by data clustering. Then the membership function shapes are learned by training on membership grades to each cluster. More detail will be given in section 2.

(b) Indirect Multi-Dimensional Membership Functions Design:
This method synthesizes multi-dimensional membership functions by combining one dimensional membership functions. The membership functions are tuned using gradient techniques which attempt to reduce the error between the desired output and actual output of the total fuzzy system.

The advantage of method (a) is that it can generate nonlinear multi-dimensional membership functions directly; there is no need to construct multi-dimensional membership functions by combining one dimensional membership functions. The advantage of method (b) is that it can be tuned by monitoring the final performance of the total fuzzy system. We review both methods in section 2.

Many of the methods that use genetic algorithms are similar in spirit to method (b); one dimensional membership functions shapes are automatically tuned using genetic algorithms. Many of these methods only consider one or two of the previously mentioned design issues. In section 3, we describe a method which decides design issues (1),(2), and (3) simultaneously.

2 Neural Network Approaches

2.1 Direct Fuzzy Partitioning of Multi-Dimensional Input Space

This approach uses neural networks to represent multi-dimensional nonlinear membership functions and is called NN-driven Fuzzy Reasoning [3, 4].

The advantage of this method is that it can generate nonlinear multi-dimensional membership functions directly. In conventional fuzzy systems, one dimensional membership functions used in the antecedent part, are independently designed and then combined to generate multi-dimensional membership functions indirectly. It can be argued that the neural network method is a more general form of the conventional fuzzy system in that the combination operations performed are absorbed by the neural network. Conventional indirect design methods have a problem when the input variables are dependent. For example, consider an air conditioner controlled by a fuzzy system that uses temperature and humidity as inputs. In conventional design methods of fuzzy systems, the membership functions of temperature and humidity are designed independently. The resulting fuzzy partitioning of the input space resembles Figure 1(a). However, when the input variables are dependent, such as temperature and humidity, fuzzy partitioning such as Figure 1(b) is more appropriate. It is very hard to construct such a nonlinear partitioning from one dimensional membership functions. Since NN-driven Fuzzy Reasoning constructs nonlinear multi-dimensional membership functions directly, it is possible to make the partitionings of Figure 1(b).

The design steps of NN-driven Fuzzy Reasoning had three steps: clustering the given training data, fuzzy partitioning the input space by neural networks, and designing the consequent part of each partitioned space.

The first step is to cluster the training data and decide the number of rules. Prior to this step, irrelevant input variables have already been eliminated using the backward elimination or information criteria methods. The backward elimination method arbitrarily eliminates one of the n input variables and trains the neural networks with $n - 1$ input variables. The performance of neural networks with n and $n - 1$ is then compared. If the performance of the $n - 1$ input networks is similar or better than the n input networks, then the eliminated input variable is considered irrelevant. Next the training data is clustered and the distribution the data is obtained. The number of clusters is the number of rules.

The second step is to decide the cluster boundaries from the cluster information obtained in step 1; the input space is partitioned and the multi-dimensional input membership functions are decided. Supervised data is provided by the membership grade of input data to the cluster that is obtained in step 1. First a neural network with n inputs and c outputs, where n is the number of relevant input variables and c is the number of clusters determined in step 1, is prepared. Training data for this network, NN_{mem} in Figure 2, is generated by from the clustering information given by step 1. Generally, each input vector is assigned to one of the clusters. The cluster assignment is combined with the input vector to form a training pattern. For example, in the case of four clusters and an input vector which belongs to cluster 2, the supervised portion of the training pattern will be $(0,1,0,0)$. In some cases, the user may intervene and manually construct the supervised portion if s/he believes an input data point should be classified differently than given by the clustering. For example, if the user believes that a data point belongs equally to class one and two, an appropriate supervised output pattern might be $(0.5,0.5,0,0)$. After training this neural network on this training data, the neural network computes the degrees to which a given input vector belongs to each cluster. Therefore, we assume that this neural network acquires the characteristics of the membership functions for all rules by learning and can generate the membership value that corresponds to any arbitrary input vector. The fuzzy systems, which uses a neural network as the membership generator is the NN-driven Fuzzy Reasoning.

The third step is the design of the consequent parts. Since we know which cluster to assign an input data to, we can train the consequent parts using the input data and the desired output. A neural network expression can be used here as in [3, 4], but any other of the proposed methods, such as math equations or fuzzy variables, can be used instead. The essential point of this model is the neural networks which partition the input space in fuzzy clusters.

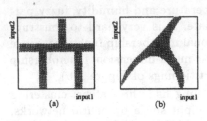

Fig. 1. Fuzzy partitioning: (a) conventional (b) desired

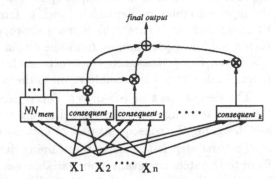

Fig. 2. Example structure of NN-driven Fuzzy Reasoning

Figure 2 shows one example of a NN-driven Fuzzy Reasoning system. This example is a model which outputs a singleton value computed by a neural network or a TSK model. Multiplication and addition in the figure calculate a weighted average. If the consequent part outputs fuzzy values, proper t-conorm and/or defuzzification operations should be used.

2.2 Tuning of Parameterized Fuzzy Systems

The parameters which define the shape of the membership functions are modified to reduce error between output of the fuzzy system and supervised data. Two methods which have been used to modify these parameters are: gradient-based methods and genetic algorithms. The genetic algorithm methods will be described in the next section and the gradient based methods will be explained in this section.

The procedure of the gradient based methods are: (1) decide how to parameterize the the shape of the membership functions (2) tune the parameters to minimize the actual output of the fuzzy system and the desired output using gradient methods, commonly steepest descent. Center position and width of the membership functions are commonly used shape definition parameters. Ichihashi et al. [6] and Nomura et al. [7, 8], Horikawa et al. [9][10], Ichihashi et al.[11] and Wang et al. [12], Jang [13][14] have used triangular, combination of sigmoidal, Gaussian, and bell shaped membership functions respectively. They tune the membership function parameters using steepest descent methods.

Figure 3 shows this method and is isomorphic to Figure 4. μ_{ij} in the figure is the membership function of input parameter x_j in the i-th rule, and actually it is represented by a parameter vector that describes the shape of the membership function. Namely, the method casts the fuzzy system as a neural network by representing membership functions as weights and rules by nodes which perform

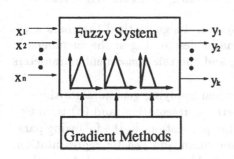

weighted average

Fig. 3. Neural networks tune the parameters of fuzzy systems

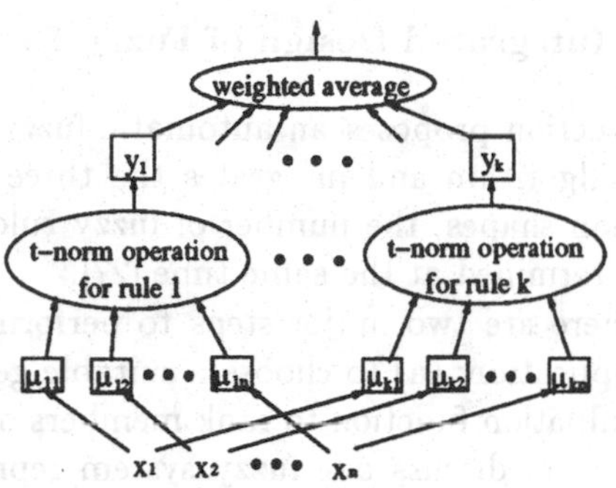

Fig. 4. Neural networks for tuning fuzzy systems: μ_{ij} is the membership function of input parameter x_j in i-th rule

t-norm operations. Any network learning algorithm, such a backpropagation, can be used to train this structure. y_i in the figure is a trainable output value of each consequent part. This method has already been applied in the design of actual commercial products.

3 Genetic Algorithm Approaches

3.1 Genetic Algorithms and Fuzzy Control

Genetic algorithms are biologically inspired optimization techniques, which operate on populations of binary coded representations and perform reproduction, crossover, and mutation on them. The right to reproduce offspring for the next generation is based on a fitness value provided by the application. Genetic algorithms are attractive because they do not require the existence of derivatives, are robust, search many points simultaneously, and are able to avoid local minima.

Several papers have proposed automatic fuzzy system design methods using genetic algorithms. Much of the work has focused on tuning membership functions [17] - [25]. Other methods use genetic algorithms to determine the number of fuzzy rules [18, 26]. In [26], sets of rules are constructed by experts and the genetic algorithm finds the best combination of them. In [18], Karr has developed a method for determining membership functions and number of fuzzy rules. In this paper, Karr's method first uses a genetic algorithm to determine the number of rules according to a predefined rule base. Following this stage, the method uses a genetic algorithm to tune the membership functions. Although these methods produce systems that perform better than human designed systems, they may be suboptimal because they treat only one or two of the three major design stages at a time. Because these design stages may not be independent, it is important to consider them simultaneously to find the optimal solution. In the next section, we propose an automatic design method that integrates the three major design stages.

3.2 Integrated Design of Fuzzy Systems using Genetic Algorithms

This section proposes an automatic fuzzy system design method that uses a genetic algorithm and integrates the three major design stages: the membership function shapes, the number of fuzzy rules, and the rule-consequent parameters are determined at the same time [27].

There are two major steps to perform when applying genetic algorithms to an application; (a) to choose a suitable genetic representation, and (b) to design an evaluation function to rank members of the population. In the following paragraphs, we discuss our fuzzy system representation and genetic representation. An evaluation function and methods for embedding apriori knowledge will be presented in the next section.

Fuzzy System and Genetic Representations We use the TSK model fuzzy system, which is widely used in control problems, to map the state of the system to a control value. The TSK model differs from conventional fuzzy systems in that the consequent parts of a TSK fuzzy model are expressed by linear equations as opposed to fuzzy linguistic expressions. For example, a rule in a TSK model has the form:

IF X_1 is A and X_2 is B THEN $y = w_1 X_1 + w_2 X_2 + w_3$,

where w_n are constants. The final control value is computed by summing the output of each rule and weighting it according to the rule's firing strength.

For the antecedent part, we use triangular membership functions parameterized by left base, right base, and distance from the previous center point (the first center is given as an absolute position). Other parameterized shapes, such as sigmoidal, Gaussian, bell, or trapezoidal can be substituted. Unlike most methods, overlap restrictions are not placed on the sets in our system and the possibility of complete overlap exists (see Figure 5).

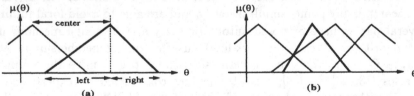

Fig. 5. (a) membership function representation (b) possible member functions

Generally, the number of fuzzy sets for each input variable combine to determine the number of fuzzy rules. For example, a TSK fuzzy system with m input variables, each partitioned into n fuzzy sets, would yield n^m fuzzy rules. Because the number of rules depends directly on the number of membership functions, eliminating membership functions has the direct effect of eliminating rules.

Each membership function requires three parameters and each fuzzy rule requires three parameters. Thus, an m-input-one-output system with n fuzzy sets per input variable requires $3(mn + n^m)$ parameters.

The genetic representation explicitly contains the three membership function parameters and the consequent parameters as mentioned previously. The number of rules, however, are encoded implicitly via the application boundary conditions and membership function positions. We can implicitly control the number of rules by eliminating membership functions whose center positions lie outside the range of its corresponding input variable and rules which contain them. For example, in the inverted pendulum application, rules using θ membership functions with center positions greater than $90°$ can be eliminated. This implies that the number of rules can be optimized at the same time the genetic algorithm optimizes the shape of the membership functions and the consequent parameters.

The actual genetic code is formulated by defining a chromosome as a set of parameters that represent a higher level entity, such as a membership function or rule-consequent parameter set (see Figure 6). Figure 7 shows chromosomes linked together to form the entire fuzzy system representation.

center	left base	right base	w_1	w_2	w_3
10100110	10011000	01011000	10100110	10011000	01011000

membership function chromosome (MFC)   rule-consequent parameters chromosome (RPC)

Fig. 6. Composite chromosomes

fuzzy variable θ			fuzzy variable $\delta\theta/\delta\tau$			rule-consequent parameters		
MFC_1	$\cdots$	MFC_{10}	MFC_1	$\cdots$	MFC_{10}	RPC_1	$\cdots$	RCP_{100}

Fig. 7. Gene map

3.3 Evaluation

Evaluation of Inverted Pendulum Controllers The inverted pendulum represents a classic non-linear control problem where the task is to find a control strategy that can balance a pole on a movable cart. In our simulation, the movement of both the pole and the cart is restricted to the vertical plane and the cart is allowed to move infinitely in either direction along the track. The controller uses the angular displacement and velocity to compute a force, which is applied throughout the control interval. More details can be found in [14] and [27].

Unlike the fuzzy system representation, the evaluation function relies directly on the application. The genetic algorithm uses an objectively computed performance measure to guide the search for better and better controllers. An inverted pendulum trial can end in three different conditions: (1) the pole becomes balanced, (2) simulation time runs out, or (3) the pole falls over. t_{end} represents the time when the trial actually ends and t_{max} represents the maximum simulation time. According to these three conditions, we came up with the following guidelines: if the system balances the pole, a shorter time is better than a longer time;

if the pole falls over, a longer time until failure is better than a shorter time. Figure 8 captures this notion.

$$score(t_{end}) = \begin{cases} a_1\,(t_{max} - t_{end}) + a_2 reward & (1) \\ reward & (2) \\ b \cdot t_{end} & (3) \end{cases}$$

where a_1, a_2, b and $reward$ are constants, and (1) pole balanced, (2) $t_{max} = t_{end}$, and (3) pole fell over ($|\theta| \geq 90°$). t_{end} represents the time when 525 the trial actually ends and t_{max} represents the maximum 526 simulation time.

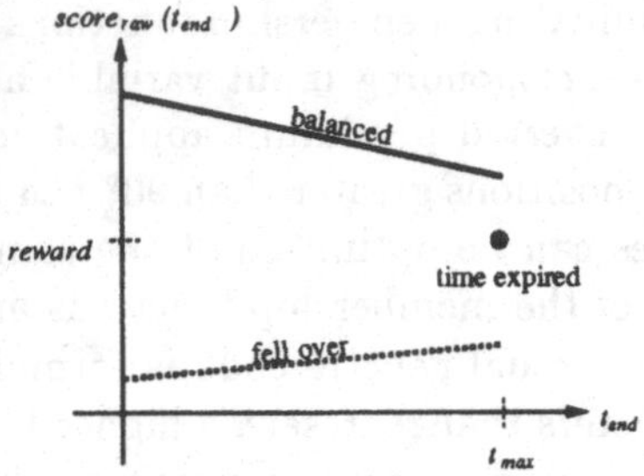

Fig. 8. Raw scoring functions

It is also desirable to have the controller work well over a wide range of initial conditions. To consider this point, we evaluated a controller by summing the scores obtained from individual trials performed on a set of initial conditions. We then augmented our evaluation function with additional terms to consider steady state error and to penalize systems according to the number of rules in the system. The resulting fitness score for one trial was computed as

$$score(t_{end}) = \frac{\left(score_{raw}(t_{end}) + c\sum_0^{t_{end}} |\theta_t|\right)}{\text{number of rules} + \text{offset}_{rules}}$$

The steady state error was a weighted integral of the angle displacement and the offset$_{rules}$ parameter controlled the degree of penalty for number of rules.

Embedded knowledge There are two levels at which apriori knowledge can be added to our fuzzy system design method: at the level of fuzzy systems and at the level of the application. We can incorporate our knowledge of fuzzy systems and inverted pendulums into our system in several ways: via initial conditions, via the representation of the fuzzy system, or via the objective function [28].

Traditional genetic algorithms start with randomly generated population of solutions. The purpose of this procedure is to randomly cover the solution space with the hope that one may be near the optimal solution. We can use our application knowledge to initialize some members of the population with good parameter sets. Biasing a population in this manner does not restrict the genetic algorithm to searching for solutions in the neighborhood of the initial solutions, because the mutation, crossover, and reproduction operations still allow the genetic algorithm to search points far from the initial points. If the initial solutions are approximately correct, we can gain significant speedup. However if they are near a local minimum, the genetic algorithm can be temporarily distracted and require more time to find the optimum.

From our experience with fuzzy systems, we know that equally partitioning the input dimensions is a good idea. Using this knowledge, we can initialize a few members of our population to cover the input dimensions in this manner. Our initial population includes members that equally divide the input dimension into varying number of fuzzy sets in addition to randomly generated members.

The inverted pendulum has been well studied and control laws such as

$$force = c_1 sin(\theta) + c_2 \frac{\delta\theta}{\delta t}$$

have been developed. We can approximate this function by setting the constant parameter, w_3, of the TSK consequent part to the value of the control law when it is evaluated at the center point of the antecedent membership functions.

We can also take advantage of the symmetrical nature of the inverted pendulum balancing task by requiring the underlying fuzzy system to be symmetric. An example symmetric fuzzy system would symmetrically partition the input space about the origin and constrain the consequent parameters to reflect symmetry. In addition to possibly improving the performance of the resulting system, the number of initial conditions to train upon can be reduced by half.

An alternative technique to incorporate symmetry into the design system is to implicitly include it into the evaluation function. By including symmetric initial starting points in the evaluation process, we can insure that symmetric conditions are considered.

3.4   Experimental Method and Results

Our method combines a genetic algorithm, a penalty strategy, and unconstrained membership function overlap to automatically design fuzzy systems. In this section, we first present results which do not make use of any apriori knowledge. We then compare these results with results obtained from our method where apriori knowledge has been incorporated into the design process.

Experimental Method   There are experimental parameters associated with the genetic algorithm and parameters associated with the application. Crossover rate, number of crossover points, mutation rate, population size, and number of generations to produce are parameters of the genetic algorithm. The maximum fuzzy sets per input variable and parameter bit resolution are parameters associated with the fuzzy system and fuzzy system coding. The offset$_{rules}$, maximum simulation time, reward, and balancing criteria are parameters used for evaluating controllers.

We used a genetic algorithm with two point crossover and set crossover rate to 0.6, the mutation rate to 0.0333, and the population size to 10. The number of generations produced depended on the experiment. We used an elitist strategy in which the member with the highest fitness value automatically advanced to the next generation.

The maximum fuzzy sets per input variable was set to ten in our experiments. This limit was set through experience with the inverted pendulum application. Because we included a penalty strategy that involves the number of rules, the setting of this number is not so critical. The precision of all parameters was set to 8 bits. The resulting genetic representation for fuzzy systems used in our experiments consisted of 360 parameters or 2880 bits (see Figure 6).

The balancing criteria was set to 0.0001 in our experiments. The pole is considered balanced if the following condition is met:
$$(\theta(t) - \theta(t-1))^2 + (\dot{\theta}(t) - \dot{\theta}(t-1))^2 < \text{balance criteria.}$$
The reward was set to 5000 and the maximum simulation time t_{end} was 200. The evaluation parameter offset$_{rules}$ was set to 10.

Experimental Results After all of these parameters have been set, the genetic algorithm begins its search. Each controller in the population was evaluated with the set of initial conditions: $(\theta, \delta\theta/\delta t) = (5.22, 6.93), (5.11, 6.97), (-8.41, -1.37), (6.22, -7.14)$. After completing the requested number of iterations, the best solution is kept and the rest are discarded. In one experiment, the method produced a system with only four rules. Their symmetric rules are:

IF θ is A_i and $\dot{\theta}$ is B_i, THEN $y = w_{1i}\theta + w_{2i}\dot{\theta} + w_{3i}$,

where $i = 1 \sim 4$. The obtained parameters in four consequent parts were (w_{1i}, w_{2i}, w_{3i}) = (0.44, 1.02, -31.65), (1.54, -0.61, -30.14), (1.54, -0.61, 30.14), and (0.44, 1.02, 31.65). The obtained triangular membership functions, A_i and B_i were $A_1 = A_3$ = {-119.65, -62.12, 4.59}, $A_2 = A_4$ = {-4.59, 62.12, 119.65}, $B_1 = B_3$ = {-219, -1.99, 238.56}, and $B_2 = B_4$ = {-238.56, 1.99, 219.64}. Figures 9 and 10 show trajectory plots for several initial conditions.

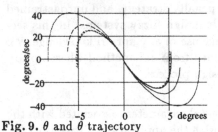

Fig. 9. θ and $\dot{\theta}$ trajectory

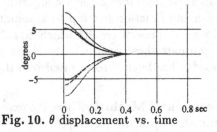

Fig. 10. θ displacement vs. time

We have designed four experiments to study the effect of including different combinations of apriori knowledge on the performance of both the design method and the resulting systems. The forms of apriori knowledge embedded into the design method were symmetric fuzzy system structure and heuristic initialization. The fitness vs generation plot appears in Figure 11 with an experimental condition table.

Each experiment performed 5000 iterations on eight symmetric initial conditions. The experiments that used symmetric fuzzy systems used only half of the points and thus performed half as many function evaluations as the asymmetric systems. It is interesting to note that while the experiments heuristically initialized initially had superior performance than those randomly initialized, they required more time to obtain higher fitness levels. This may be due to the fact that the initial conditions given represented a local minimum.

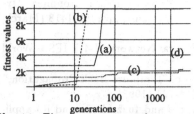

Fuzzy rules and initialization of genetic algorithm parameters

experiment	Fuzzy System	Initialization
(a)	symmetry rules	heuristic
(b)	symmetry rules	random
(c)	asymmetry rules	heuristic
(d)	asymmetry rules	random

Fig. 11. Fitness vs. generation

4 Conclusion

Neural networks and genetic algorithms have matured into practical tools for automatic fuzzy system design. Neural network approaches for automatically and completely specifying fuzzy systems have already made their commercial debut. Neural networks and genetic algorithms for automatic fuzzy system design are just two examples of using one technology to strengthen another. Many more combinations involving two or more cooperating technologies are waiting to be discovered and explored.

References

1. Takagi, T. and Sugeno, M., "Fuzzy Identification of Systems and Its Applications to Modeling and Control," IEEE Trans. SMC-15-1, 1985, pp.116-132
2. Mamdani, E. H., "Applications of fuzzy algorithms for control of simple dynamic plant," Proc. of IEEE, Vol. 121, No. 12, pp.1585-1588 (1974)
3. Takagi, H. and Hayashi, I., "Artificial_neural_network-driven fuzzy reasoning," Int'l Workshop on Fuzzy System Applications, pp.217-218 (Aug., 1988)
4. Takagi, H. and Hayashi, I., "NN-driven Fuzzy Reasoning," Int'l J. of approximate Reasoning, Vol. 5, No.3, pp.191-212 (1991)
5. Smith, M.H., "Parallel Dynamic Switching of Reasoning Methods in a Fuzzy System," 2nd IEEE Int'l Conf. on Fuzzy Systems, vol.2, pp.968-973 (March, 1993)
6. Ichihashi, H. and Watanabe, T., "Learning Control by Fuzzy Models Using a Simplified Fuzzy Reasoning", J. of Japan Society for Fuzzy Theory and Systems. Vol. 2, No.3, pp.429-437 (1990), (in Japanese)
7. Nomura, H. Hayashi, I. and Wakami, N., "A self-tuning method of fuzzy control by descent method," 4th IFSA World Congress, Vol. Engineering, pp.155-158 (July, 1991)
8. Nomura, H. Hayashi, I. and Wakami, N., "A learning method of fuzzy inference rules by descent method," IEEE Int'l Conf. on Fuzzy System, pp.203-210 (March, 1992)
9. Horikawa, S., Furuhashi, T., Okuma, S., and Uchikawa, Y., "Composition Methods of Fuzzy Neural Networks," Int'l Conf. on Ind., Elect., Control, Instr., and Automation , pp.1253-1258 (Nov., 1990)
10. Horikawa, S., Furuhashi, T., and Uchikawa, Y. "On Fuzzy Modeling Using Fuzzy Neural Networks with the Back-Propagation Algorithm," IEEE Trans. Neural Networks. Vol.3, No.5, pp.801-806 (1992)
11. Ichihashi H. and Tanaka, "Backpropagation Error Learning in Hierarchical Fuzzy Models," Symposium of SICE Kansai Chapter, pp.131-136 (1990) (in Japanese)

12. Wang, L-X. and Mendel, J.M., "Back-Propagation Fuzzy System as Nonlinear Dynamic System Identifier," IEEE Int'l Conf. on Fuzzy Systems, pp.1409-1418 (March, 1992)

13. Jang, J-S. "Rule Extraction Using Generalized Neural Networks," 4th IFSA World Congress. Vol.Artificial-Intelligent, pp.82-86 (July, 1991)

14. Jang, J-S. "Self-Learning Fuzzy Controllers Based on Temporal Back Propagation," IEEE Trans. Neural Networks. Vol.3, No.5, pp.714-723 (1992)

15. Takagi, H., "Cooperative system of neural networks and fuzzy logic and its application to consumer products," (edited by J. Yen and R. Langari) Industrial Applications of Fuzzy Control and Intelligent Systems, Van Nostrand Reinhold (will be published in 1993)

16. Asakawa, K. and Takagi, H., "Neural Networks Applications in Japan," Communications of ACM, (submitted)

17. Karr, C., Freeman, L., Meredith, D., "Improved Fuzzy Process Control of Spacecraft Autonomous Rendezvous Using a Genetic Algorithm," SPIE Conf. on Intelligent Control and Adaptive Systems, pp.274-283 (Nov., 1989)

18. Karr, C., "Applying Genetics to Fuzzy Logic," AI Expert, Vol.6, No.2, pp.26-33 (1991)

19. Karr, C., "Design of an Adaptive Fuzzy Logic Controller using a Genetic Algorithm," Int'l Conf. of Genetic Algorithms, pp.450-457 (July, 1991)

20. Karr, C., and Gentry, E., "A Genetics-Based Adaptive pH Fuzzy Logic Controller," Int'l Fuzzy Systems and Intelligent Control Conf., pp.255-264 (March, 1992)

21. Karr, C., Sharma, S., Hatcher, W., and Harper, T., "Control of an Exothermic Chemical Reaction using Fuzzy Logic and Genetic Algorithms," Int'l Fuzzy Systems and Intelligent Control Conf., pp.246-254 (March, 1992)

22. Nishiyama, T., Takagi, T., Yager, R., and Nakanishi, S., "Automatic Generation of Fuzzy Inference Rules by Genetic Algorithm," 8th Fuzzy System Symposium, pp.237-240 (May, 1992) (in Japanese)

23. Nomura, H., Hayashi, I., and Wakami, N., "A Self-Tuning Method of Fuzzy Reasoning By Genetic Algorithm," Int'l Fuzzy Systems and Intelligent Control Conf., pp.236-245 (March, 1992)

24. Qian, Y., Tessier, P., and Dumont, G., "Fuzzy Logic Based Modeling and Optimization," 2nd Int'l. Conf. on Fuzzy Logic and Neural Networks , pp.349-352 (July, 1992)

25. Tsuchiya, T., Matsubara, Y., and Nagamachi, M., "A Learning Fuzzy Rule Parameters Using Genetic Algorithm," 8th Fuzzy System Symposium, pp.245-248 (March, 1992) (in Japanese)

26. Takahama, T., Miyamoto, S., Ogura, H., and Nakamura, M., "Acquisition of Fuzzy Control Rules by Genetic Algorithm," 8th Fuzzy System Symposium, pp.241-244 (March, 1992) (in Japanese)

27. Lee, M., and Takagi, H. "Integrating Design Stages of Fuzzy Systems using Genetic Algorithms," 2nd IEEE Int'l Conf. on Fuzzy Systems, vol.1, pp.612-617 (March, 1993)

28. Lee, M., and Takagi, H. "Embedding Apriori Knowledge into an Integrated Fuzzy System Design Method Based on Genetic Algorithms," 5th IFSA World Congress (1993) (to appear)

29. Goldberg, D., "Genetic Algorithms in Search, Optimization, and Machine Learning," Addison-Wesley (1989)

SYMBOLIC AND NUMERIC DATA MANAGEMENT IN A GEOGRAPHICAL INFORMATION SYSTEM: A FUZZY NEURAL NETWORK APPROACH

ZAHZAH E.H., DESACHY J.

Université Paul Sabatier
Institut de Recherche en Informatique de Toulouse
118 Route de Narbonne 31062 Toulouse Cedex FRANCE

☎ :(+33) (61 55 65 99)
Fax :(+33) 61 55 62 58)
E-mail: zahzah@irit.fr, desachy@irit.fr

Abstract. The problem of extracting information issued from several sources of information turns out to be a very important issue in intelligent systems. This problem is always encountered in multi-expert systems.
In the field of remote sensing and geographic information system, this question is well known. Satellite images, geographic and geologic data, and expert knowledge can appear independent but, when pooled together, they can give more informations about a same object or a same problem than used separately. This data may be very heterogeneous(from simple numeric items to complex symbolic information).
In this paper we propose a common scheme to combine numeric and symbolic information by means of fuzzy neural networks techniques. As an application, we describe a method for complex geographical information extraction based on standard geographical information and expert symbolic knowledge .

1 Introduction

Nowadays most cartographic applications manage with satellite imagery, but almost all of them limit themselves to in-image information (spectral or textural features, multitemporal images...). Actually when a photointerpreter analyzes a satellite image, he takes into account a lot of "a priori", "out-image" knowledge to reach a satisfactory interpretation. Λ photointerpreter who wants to produce a vegetation map for example, will manage with three types of information sources.
a) satellite image,
b) available cartographic information (topography, soils quality etc...),
c) his knowledge of local vegetation types characteristics.

So it's quite an evidence that one should "add" to image information, the domain expert knowledge "know-how" and the cartographic data in order to "understand" the image.

Our team has already conceived an expert system ICARE (Image CARtography Expert) based on production rules and certainty factors. We give a quick description of this system in order to introduce the basic concepts of our approach.

The aim of the system is to improve usual supervised classification used to produce "maps" from remote sensing imagery[2]. The expert knowledge is stored in a knowledge base as a set of production rules with certainty factors. The fact data base is a set containing the image to be interpreted and the associated geographic information system.

2 Expert system description

The general scheme of the system is presented in fig 1., let's detail each part .

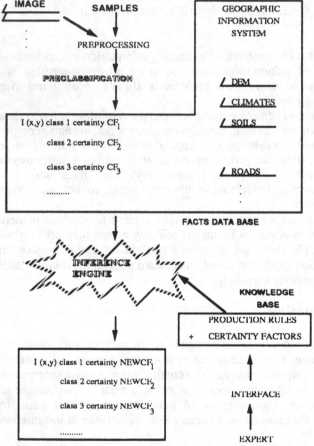

Fig.1. General Scheme of the expert System (ICARE)

2.1 The Facts Data Base

An elementary fact in the system is the whole set of available informations concerning each pixel (Some of them may not be immediately available but can be computed). The preclassification (maximum likelihood classifier) is assumed to be already realized. So a fact appears as follows .

a)	(x,y)	Pixel Coordinates
	Class 1	Certainty Factor CF1
	Class 2	Certainty Factor CF2
	.....	....
	Class n	Certainty Factor CFn

CFi is the certainty factor for the pixel (x,y) in the image to belong to the class i.

b) Pixel 's context: (Known or to be computed using the Geographic Information System), related to the Digital Elevation Model: Ridges, bottom of Valley, Slope, Aspect etc...or related to Soil.

2.2 The Knowledge Base

The expert (Photointerpreter) expresses his knowledge in two steps and in two different ways.

First, he points out in the image areas which are samples of all classes. These prototypes will be used to find the class distribution model. Actually, the preclassification is the step where for each element one associated the confidence measurement that an element belongs to each class. This information is given for example by the posterior probability P(Ci/x) calculus.

At the second step, a second type of knowledge is given (symbolic) which is based on the experience and the "Know-How" of the expert, who defines for example relationships between classes and their geographical contexts as:

"Pines are principally located on south aspectss from 800m to 1500m."

This knowledge links the class pines to its context, this symbolic information is combined to the preclassification result in order to improve the classification. Note that this knowledge is fuzzy, as it is expressed in a pseudo-liguistic form, it also introduces uncertainty (Principally) and imprecision (south aspects from 800m to 1500m). The fuzzy knowledge will be mapped to IF...THEN... fuzzy rule production and the set of all the production rules related to all classes will constitute the knowledge data base

2.2.1 Knowledge description

The system is conceived for automatic cartography, the expert's knowledge we are managing with is related to classification problems, it means that this knowledge concerns the classes we are looking for in the image, and handles information existing or computable in the geographic information system or eventually on the image itself.

The system manages with simple and complex data. Simple data means that data is extracted easily by image reading (radiometric responses from satellite image), or by performing simple processes on it (slope and aspect images from the Digital Elevation Model image). Complex data means that informations are obtained by quite more complex processes involving fuzzy symbolic knowledge. For our example

> "Pines are principally located on south aspects from 800m to 1500m."

- "Principally", is an adverb which shows uncertainty on the knowledge.
- "south aspects", concerns the aspect which may be deduced from the digital elevation model.
- "from 800m to 1500m", concerns the digital elevation model directly.
This knowledge unit will be translated in the following production rule in the ICARE system

> **IF** (class pines) **THEN** (elevation from 800m to 1500m) AND (south aspects)

To the adverb "principally" corresponds a value of certainty factor of the rule, which estimates sort of a frequency degree of the class pines in the given context (-1.0 means 0% , 0.0 means 50%, +1.0 means 100%). The expert expresses his knowledge with predefined key-words :
- Key-words related to objects that can be extracted from the geographic information system(Valley, aspect, ridge, plateau, irrigated zone, road, village, clay, rocky ...etc).
- Key-words for relation capabilities: Far away, near, around ...etc.
-Key-words for uncertainty expressions Always (+1.0), often(+0.8),..., seldom(-0.8), never(-1.0).

2.3 Inference Engine

Applying productions rules of the knowledge base to the primary facts data base (result of a supervised classification + GIS information), we obtain a final fact data base (Final classification) as shown in Fig.1.
The inference engine will proceed by "Forward chaining": for each elementary fact, all production rules will be activated since each element(pixel) belongs to each class with different certainty factors. So for each pixel and each possible class, we have the following process:

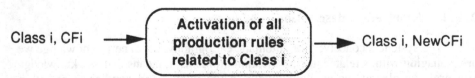

The certainty factors CFi, are updated using Mycin like uncertainty management. (see for more details [2]). Note that the premises of a rule can be in the form of combination of unions and intersections, in this case the following computations are assumed to be done previously.

IF Classi THEN (A∪B) → CF(A∪B) = Max(CFA,CFB)

IF Classi THEN (A∩B) → CF(A∩B) = Min(CFA,CFB)

At the end of the process, each point is assigned to each class with updated certainty factors, and the final classification can take place.

In the following section, The ICARE Expert System approach is used to describe a method for complex geographical information computation based on standard geographical information and expert symbolic knowledge by fuzzy neural networks techniques. The final result will be a potentiality map giving a realization degree of the corresponding rule (Problem) for each point. That is linked to ICARE system approach for knowledge representation and combination of various sources of information, but here no remotely sensed image information is involved. Moreover one of the main disadvantages of the rule based approach is that each classification processing induces activation of all rules for all pixels in the image which produces 10 minutes CPU time on a SUN SPARC for our application. So we have introduced the fuzzy neural networks approach to speed up the computations.

3 Fuzzy Neural Networks for Knowledge Representation

Let's give for example a problem of information extraction related to an irrigated zone. An expert can define such zones, as regions which are "always near hydrographic network and frequently less than 300m and often near roads" for example. This knowledge is represented by a fuzzy production rule as defined above.

IF(Irrigated zone) **Then** (always near hydrographic network) And (frequently less than 300m) And (often near road).

One assumes that a rule corresponds to a particular situation and describes the ideal context for realization of a complex situation with frequency degrees, and is represented by a neural network [8]. The net inputs are the realization degrees of all possible atomic propositions expressed in the conclusion part of the rule , "near a road", "elevation less than 300m", "near hydrographic network", (Fig.2.)

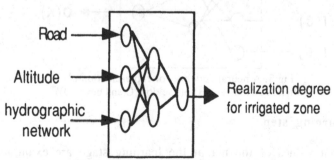

Fig.2. Knowledge Unit Representation

These inputs take values in the interval [-1.0,+1.0], the value -1.0 corresponds to the case of false/absent proposition, +1.0 when it is true/present, and 0.0 to the case of unknown/no information, these values are modulated by the frequency degrees associated globally to the rule or to each atomic proposition.

A knowledge unit R is considered as a combination of disjunctive and conjunctive propositions as one knows how to map a complex rule into a simple expression with no redundancy propositions [7,8]

$$R = \bigcup_{i=1}^{n} P_i \qquad P_i = \bigcap_{j=1}^{m} p_{i,j}$$

$$p_{i,j} = <c\,f_r>_{p_{i,j}} \quad <c\,f>_{p_{i,j}}$$

pi,j is an atomic premise, and is expressed by a frequency degree given to the premise itself(cfr), followed by a realization degree of the premise(cf). The knowledge unit R is realized, if at least one of its disjunctive proposition Pi is realized. At the learning stage, the net learns how each disjunctive proposition and its negation is realized .

Actually, learning R consists in the following process :
when inputs are in the context of R, the output is set to the maximum value of realization of the rule, when inputs are in a wrong context of R, the output is set to the minimum value of realization of the rule , and intermediate values are output for intermediate situations.[7,8]
example:
R: IF "favourable context" THEN (often A AND always B) OR (seldom C AND never D)

This rule (Fig.3.) is defined by four atomic propositions (A, B, C, D), the adverbs "often" and "always " tell us about the frequency to be in the context of A (resp B), "often" means that about 80% of favourable contexts are in the context of the premise A, "always" means that 100% of them are in the context of the premise B. The same reasoning is done with (seldom C and never D)

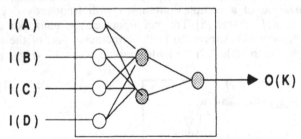

Fig 3. A neural network for representation of
((often A and always B) or (seldom C and never D))

3.1 The learning step

The input values of the net at the learning stage are examples in the outstanding cases (Fig.4.) , where each premise may be totally realized or not. The net produces its inner representation and hence, can generalize and resolve the intermediate situations. At this level, the adverbs are temporarily ignored .

The net is hierarchical networks of nodes, that contain processing units which perform a memoryless nonlinear transformation on the weighted sum of their inputs. A node produces a continuous-valued output between -1.0 and 1.0. Weights are positive or negative real values determined during training. The output O(k) gives the realization degree of the rule belonging to the interval [+1.0 - 1.0], +1.0 corresponds to the maximum value of the realization of the rule, -1.0 to the minimum value, and the value 0 corresponds to the case of total ignorance. The net uses the backpropagation learning algorithm [4].

I(A)	I(B)	O(K)
1	1	1
1	0	α
1	-1	β
0	0	0
0	1	α
0	-1	χ
-1	1	β
-1	0	χ
-1	-1	-1

Fig. 4. input/output for learning the rule
if "favourable context" then $(A \cap B)$
$O(k) = f(I(A), I(B))$, $-1.0 < \chi \leq 0 < \beta < \alpha < 1.0$

For instance , we learn the net that, when all the premises of a given rule are realized (all the correspondent inputs are set to 1.0), then the output is then set to 1.0 (maximum degree of realization of the concerned rule), when all the premises are impossible (all the correspondent inputs are set to -1.0) the output is set to -1.0 (minimum degree of realization of the corresponding rule), and when half of the rule's premises are realized and the others are not, or no premise at all is realized, then the net learns that the situation is more favourable for the first case than in the second one, and must output values which are relatively great and little for the two respective cases. Geometrically, the system simulates the behaviour of the function of Fig5. For a multidimensional system, the net resolves the system :

$$[-1.0 +1.0]^n \; --o--> \; [-1.0 +1.0] \quad , \quad O(k) = f(I(P1), I(P2), ..., I(Pn))$$

where Pi are premises, I(Pi) are certainty factors and O(k) is the realisation degree of the context defined by the knowledge unit.

When learning "if favourable context then (A and B)", the inputs for the premises C and D are disactivated, and when learning "if favourable context then (C and D) " the inputs for A and B are disactivated. The disactivation of an input is performed by assigning a zero value to the corresponding value for cf_r.

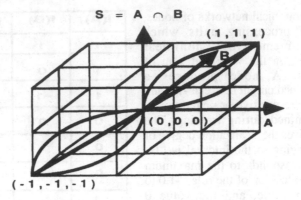

Fig. 5 The system's behaviour for the recognition of a rule
S=A ∩ B (A and B are any given proposition).

3.2   System operating

At this stage, the system is functional and ready to respond to any inputs, the output tells us wether the inputs correspond or not to the rule, and how well they do. Note that he adverbs preceding the premises in the rule are introduced at this stage. Effective inputs for the net will be the product of certainty factors of the premises, with their corresponding frequency degrees (0.8 for "often, +1.0 for always etc..). At the learning stage, the rule was learned with frequency degrees equal to 1.0, as if the premise 's rule were all defined by "always". At this stage, these frequency values are taken into account in order to be conform to the knowledge as it has been expressed by the expert [8]. After the presentation of the pixel's context to the net , the system outputs the realization degrees corresponding to the rule it represents.

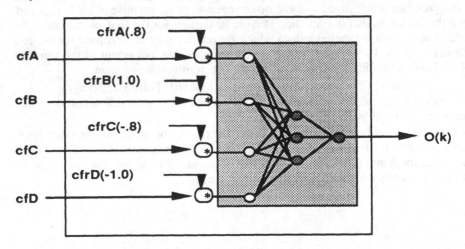

Fig6. net representation of knowledge:
(often A and always B) or (seldom C and never D)

4   Applications

We applied this method to extract complex information (Irrigated Zone) described by fuzzy symbolic knowledge from Digital Elevation Model (DEM), derived information (hydrographic network map) and roads map. We tested the method with several definitions of irrigated zones, by changing adverbs.

image1 corresponds to the DEM image, it is represented in four elevation levels, white regions are the highest zones, and dark regions are the lowest zones.
image2 : roads map
image3 : hydrographic network

Image4 represents the result by applying a first definition of "irrigated zone " to the system:
IF(Irrigated zone) Then
     (always near hydrographic network) AND
     (frequently less than 300m) AND
     (often near road).
Actually, the resulting image is a set of points in the interval of [-1.0,+1.0], image4 , image5, image6 are representations by classes. White regions are the most favourable zones to the corresponding definition, grey regions are less favourable, and so on.

Image 5 represents the result by applying a second definition given by the following rule:
IF(Irrigated zone) Then
     (often near hydrographic network) AND
     (often less than 300m) AND
     (often near road).

Image 6 is the result by using the third definition
IF(Irrigated zone) Then
     (always near hydrographic network) AND
     (frequently less than 300m) AND
     (never near road).

We can notice from the successive images, the changes of patterns with the modification of definitions, favourable zones increase when adverbs are less restrictive , and decrease when adverbs are more restrictive.

5   Conclusion

We have applied this fuzzy neural net technique to represent expert knowledge for complex premises computation in the frame of ICARE expert system. This technique can be used to produce potentiality map for any given realistic problem , as "Where rice can be cultivated?" "What are the exposed regions to fire ?" or " What are the exposed regions to erosion ?"
It may also be introduced in the frame of geographical information systems in order to produce potentiality maps for problem solving.

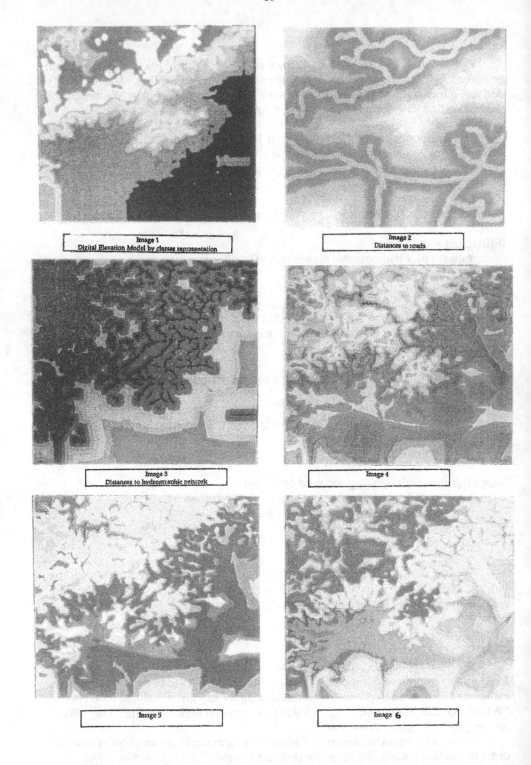

Image 1
Digital Elevation Model by classes representation

Image 2
Distances to roads

Image 3
Distances to hydrographic network

Image 4

Image 5

Image 6

References & Bibliography

1.   DESACHY J. , Debord P. , Castan S."An expert system for satellite image interpretation and GIS based problem solving"16 th international congress of ISPRS Kyoto   (1988)

2.   DESACHY J "I.C.A.R.E. : An expert system for automatic mapping from satellite imagery" in NATO-ASI series vol F6 "Mapping and spatial modelling for navigation" edited by L.F. Pau Springer verlag Berlin heidelberg   (1990)

3.   GALLANT S.I "Connectionist expert systems" , communications of the ACM, feb 1988 vol 31 n 2 pp 152-169

4.   RUMELHART D.E, Hinton G.E, Williams R.j "learning internal representation by error propagation"PDP, vol1(ed. J.L McClelland, D.E Rumelhart), MIT Press, 1986, 318-364.

5.   SANCHEZ Elie "Fuzzy connectionist expert system"Proc of ICFL &NN (Iuzuka, Japan, July 1990 pp. 31-35.

6.   WITHBREAD P.J , R.E BOGNER" "Neural networks for the recognition of multi-spectral texture" IEEE International conference on image processing ICIP'89 Singapore 5-8 septembre 1989.

7.   ZAHZAH E.H. , DESACHY J., ZEHANA M."A fuzzy connectionist knowledge based image interpretation system" 2nd International conference on fuzzy logic and neural networks - Iizuka'92 , July 92, Fukuoka Japan .

8.   ZAHZAH E-H "Contribution a la représentation des connaissances et à leur utilisation pour l'interprétation des images satellite", Thèse de l'UPS Toulouse Septembre 1992.

Approximate Reasoning in the Modeling of Consensus in Group Decisions

Luisa Mich, Mario Fedrizzi, Loris Gaio

Università di Trento, Dipartimento di Informatica e Studi Aziendali,
Via Inama, 5 38100 Trento, Italia

Abstract. In this paper we propose an approach to consensus reaching based on linguistically expressed individual opinions and on so-called opinion changing aversion. We operate within this basic context: there is a group of experts which must choose a preferred alternative from a finite set of admissible ones according to several criteria. Each expert is called upon evaluate linguistically the alternatives in terms of their performance with respect to each criterion. The task of the experts is to reach some agreement during a consensus reaching process directed by a third person called the *moderator*. The experts are expected subsequently to change their testimonies until sufficient agreement (*consensus*) has been reached. The measure of consensus depends on a function estimated for each expert according to his/her aversion to opinion change.

1 Introduction

Much of decision making activity in reality proceeds within a group framework, and it is quite natural that some computerized decision support systems can be used to help group decision making processes. Indeed group decision support systems (GDSS) are widely advocated, developed and employed both by the public administration and by private companies. For a review, see e.g., [DESA87],[EOM 90],[GRAY87],[HUBE84],[JACO92],[KRAE88],[NUNA91] and [TAKA92].

Group decision making involves choice processes whose purpose is to find an option (or a set of options) which is considered *best* by a certain group of experts (viewed as a single being) according to a set (generally finite) of criteria. The point of departure is these experts' testimonies given in the form of, say, subjective probabilities, orderings, preferences, etc. over a set of options. Consensus is a major goal of group decision making (see, e.g., [KOWI80]). By, *consensus* is traditionally meant a full and unanimous agreement (concerning the above mentioned individual testimonies), although this is obviously by no means a well-defined and clear-cut definition (see, e.g., practically all the papers in [LOEW85]). In any non-trivial case the experts initially disagree over their testimonies, so that consensus is to be viewed within a dynamic perspective. That is, assuming that the experts are seriously committed to consensus, via exchange of information and rational argument they update step by step their testimonies, which come increasingly to resemble to each other, and - hopefully - consensus in the sense of unanimous agreement is finally obtained.

Unfortunately humans rarely reach such that unanimous agreement, and even if this were the case, the consensus reaching process could be too long for practical purposes. This entails reconsideration of both the essence of consensus and the consensus reaching process. We may view consensus as not necessarily full and unanimous agreement; for instance [LEVI74], operating in the context of subjective probabilities, suggests viewing consensus not as a single probability assignment, but as a set of them. Moreover, we may aknowledge that experts, though seriously committed to rational consensus, are not willing fully to change their testimonies, so that consensus is not unanimous agreement, i.e. the same testimony for all, but some set of individual testimonies which are sufficiently similar.

In this paper, according to the general framework developed in [BUI 87], [CARL92],[FEDR88],[FEDR91] and [KACP88], we propose a slightly different approach to consensus reaching based on linguistically expressed individual opinions and on so-called opinion changing aversion. We operate within the following basic context. There is a group of experts which must choose a preferred alternative from a finite set of admissible ones according to several (finite) criteria. These experts are able to express their evaluations in a very natural way by using a limited vocabulary of linguistic terms. Thus each expert is called upon evaluate linguistically the alternatives in terms of their performance with respect to each criterion. The task of the experts is to reach some agreement during a consensus reaching process directed by a third person called the *moderator*. The experts are expected subsequently to change their testimonies - via mutual interactions such as exchange of information, negotiation, bargaining, etc., and perhaps with help from the moderator - until sufficient agreement (*consensus*) has been reached. The measure of consensus depends on a function estimated for each expert according to his/her aversion to opinion change. It aims therefore to support the moderator while monitoring and controlling this process; i.e. gauging the process towards consensus and telling him or her when to stop (there is only a finite time for the process, for practical reasons).

2 Formulation of the Problem

Let us assume that a group of experts has either to choose a preferred alternative from a set of admissible ones or to give a ranking to the set of alternatives, both operations according to several explicit criteria. In formal terms, the space of the problem is defined by means of the following sets:

- admissible alternatives:

$$A = \{A_1, A_2, \ldots, A_N\}$$

- judgement criteria:

$$C = \{C_1, C_2, \ldots, C_M\}$$

- experts consulted:

$$E = \{E_1, E_2, \ldots, E_K\}.$$

It is assumed that A and C have been defined in a previous phase of the problem definition.

A classic example of this situation is a public administration body (a group of public decision-makers) which must deal with a problem of Environmental Impact Assessment [LEE 83]. Let us image that a decision must be taken concerning the creation of a nature park. Assessment of environmental policies should take account of the conflicts arising from the trade-off between protection of the area and pre-existing production activities. In [BOAT91] a set of five hypotheses summarize the different positions of the organisations and the categories involved in the creation of a regional park in the wetlands of the Caorle Lagoon (Venice). These are:

A_1 Maintain the status quo;
A_2 Create a park only in the wetland area that is public property and a pre-park area;
A_3 Create a park in the whole wet area and an area of pre-park including a stretch of cultivated land;
A_4 The same as A_3 but permitting more activities both in the park and pre-park areas;
A_5 Create the park in all areas of natural interest in eastern Veneto with more constraints than in A_4.

The experts are asked to evaluate the impact of each hypothesis (alternative) on the natural and socio-economic environment on the basis of different factors (criteria): e.g. the socio-economic costs to the various groups affected by the creation of a protected area (tourist operators, farmers, fish-farmers, etc.) in terms of employment, value added, stock capital; benefits for the natural environment, management costs of the projects, etc.

To enable the experts to express their opinions in a most natural way, the system provides them with a limited vocabulary of linguistic terms. That is to say, the experts choose a linguistic label in a term set V, the range of which is pre-established. For example we can have, $V = \{$very low,low,medium,high,very high$\}$. A linguistic label is a value for a so-called linguistic variable, a variable whose values are not numbers but words in a natural or artificial language (see [ZADE73],[ZIMM91]).

Thus each expert is called upon to evaluate the decisional alternatives in terms of their *performance* with respect to each criterion. From a formal point of view, the results of this first phase of consultation can be expressed in a matrix form. The elements of these matrices represent the *linguistic performances* that the experts have assigned to each alternative with respect to each one of the criteria. To each expert will correspond for example, a matrix of the kind set out in Tab.1.

Table 1. example matrix for Alternatives/Criteria

	C_1	C_2	...	C_j	...	C_M
A_1	high	very low				low
A_2	low	low				medium
...						
A_i	high	medium				medium
...						
A_N	very high	low				low

At the beginning of the session we have to determine the semantics of the terms in the vocabulary; that is, the meaning of *high* or *medium* etc. performance must be defined. The values for V, following Zadeh's approach [ZADE75], are represented by *fuzzy numbers* of the trapezoidal type, which are fuzzy sets defined by quadruples like the following:

$$\mu_{ij}^k = \alpha_{ij}^k, \beta_{ij}^k, \gamma_{ij}^k, \delta_{ij}^k \quad i=1,...,N; j=1,...,M; k=1,...,K. \tag{1}$$

The relative geometrical representation of this fuzzy number is given in Fig.1.

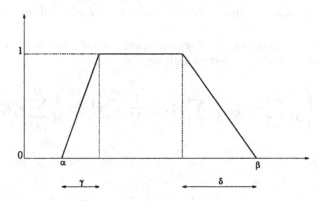

Fig. 1. plot of a trapezoidal fuzzy number

They are easy to compute and are able to capture the fuzzy uncertainties of human intuition. Much has been written in the literature about the use of fuzzy numbers. However, most authors acknowledge that they are particularly useful in those situations in which probability or utility values cannot be precisely defined but are obtained through verbal statements (see [DUBO82][DUBO87]).

3 Aggregation Method and Linguistic Approximation

We subsequently perform some classic steps in aggregation [TONG80] and linguistic approximation [FEDR92], in order to evaluate an overall linguistic value of performance for each alternative and for each expert; that is, the overall performance obtained for each expert when all the criteria are taken into account.

As a first step we must solve a multi-criteria ranking problem for each expert. In similar situations in the literature, this problem has been resolved using different approaches. One of the methods most frequently used is the so-called fuzzy outranking relation characterized by a degree of credibility which depends on two other indexes, a confidence index and a doubt index. This approach was proposed by B. Roy [ROY 77] for those situations where the elements of the matrix are punctual evaluations. Since then it as received several extensions, of which the one proposed by J.M. Martel [MART86] is of particular interest because it takes distributive evaluations into consideration.

The method we propose is closer to the one suggested by R.M. Tong and P. Bonissone [TONG80],[BONI82] who present a multi-choice model based on fuzzy information expressed linguistically. This model aggregates the linguistic performance values assigned to each alternative.

We proceed in two stages. First, we compute an average, maybe a weighted one, for the performance values. We then solve the problem of linguistic approximation.

The weighted average of the linguistic performance values is calculated for each alternative and each expert as follows:

$$\mu_i^k = \left(\frac{1}{M} \sum_{j=1}^{M} \pi_j^k \alpha_{ij}^k, \frac{1}{M} \sum_{j=1}^{M} \pi_j^k \beta_{ij}^k, \frac{1}{M} \sum_{j=1}^{M} \pi_j^k \gamma_{ij}^k, \frac{1}{M} \sum_{j=1}^{M} \pi_j^k \delta_{ij}^k \right) \qquad (2)$$

where

$$w_j^k = \frac{\pi_j^k}{\sum_{j=1}^{M} \pi_j^k} \quad \text{with } \pi_j^k = \text{weight assigned to criterion j by expert k.} \qquad (3)$$

The result of the aggregation must be translated back into linguistic terms. In fact, the fuzzy numbers obtained in the previous aggregation phase usually do not correspond exactly to terms in the initial vocabulary. We must therefore identify one term of the linguistic vocabulary and associate it with each fuzzy set of the matrix previously computed. That is, we inspect the term set and choose the linguistic label the meaning of which comes closest to the meaning of each fuzzy set corresponding to the overall performance of the alternatives.

When we have trapezoidal membership functions, the quadruples $(\alpha, \beta, \gamma, \delta)$ produce a complete representation of the fuzzy set. In order to select a label for these fuzzy sets we use the distance defined in terms of

$$d(V_l, \mu_i^k) = \sqrt{P_1^2(\alpha_l - \alpha_i)^2 + P_2^2(\beta_l - \beta_i)^2 + P_3^2(\gamma_l - \gamma_i)^2 + P_4^2(\delta_l - \delta_i)^2} \quad (4)$$

where $V_l = (\alpha_l, \beta_l, \gamma_l, \delta_l) \in V$ and $\mu_i^k = (\alpha_i^k, \beta_i^k, \gamma_i^k, \delta_i^k)$

P_1, P_2, P_3, P_4 importance of $\alpha, \beta, \gamma, \delta$ in representing a fuzzy set.

We choose the V_l^* such that

$$d(V_l^*, \mu_i^k) \leq d(V_l, \mu_i^k) \quad \forall V_l \in V, \qquad (|V| = L < \infty). \quad (5)$$

Information available after the process of linguistic approximation is represented by a bidimensional matrix $N \cdot K$ where the element in position (i, k) is the trapezoidal label of the overall linguistic value of performance attributed by the expert E_k to alternative A_i.

4 The Session for Consensus Reaching

By way of summary, the group of experts could be depicted as a black box which receives information about the problem as input and, as output, should give a satisfactory solution according to the model proposed in this paper. First of all, a suitable representation was given to the decisional problem (i.e. the possible solutions or alternatives and evaluation factors for criteria). Then the input data, i.e. the experts' individual opinions, was gathered (performances of alternatives with respect to each criteria).

After this processing of the initial data (aggregation and linguistic approximation), a *consensus degree* (see formulas (7)-(9)) on each alternative is evaluated. If consensus is reached, we exit successfully. Otherwise some of the opinions given by the experts need to be changed. In other words, if there are alternatives on which agreement has not been reached, the moderator - who could theoretically be one of the modules of the SW system - invites the experts concerned to modify their positions, taking into account their aversion to opinion change. The model provides various criteria with which to identify some consensus strategies and leads the consensus reaching process without forcing it. This is done by informing how to change these opinions in order to improve the degree of consensus. In this sense, the negotiation involves learning.

The logical architecture of the proposed model is set out in Fig.2. This graph shows clearly that there are two feedback levels in the decision: one as global evaluation, and the other as input values. In reality it would be necessary to consider a third feedback level, on the parameters of the problems, that is, on the generation of the alternatives (problem restructuring, see: [SHAK91]). To sum up, three phases can be recognized in the decisional process:

- Problem Representation
- Data Input and Evaluation of Group Opinion
- Consensus Evaluation and Changing of Opinions.

Compared to the model proposed in [DESA87], these three components correspond respectively to Input, Conflict Interaction Process and Output.

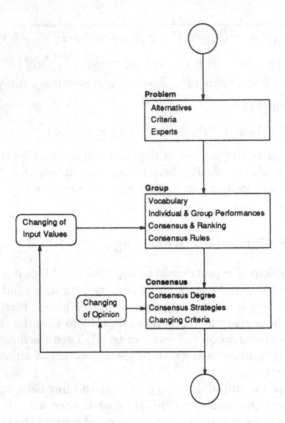

Fig. 2. logical architecture of the model

4.1 Consensus Evaluation and Changing of Opinions

After the aggregation and approximation stages, one may establish the ranking of the alternatives for each expert by using one of the many algorithms proposed in the literature (see, e.g., [BORT85] and [SAAD92]).

The last and fundamental problem to be solved is determining a *consensual ranking*. We propose a fuzzy logic based approach flexible enough not to require any particular hypothesis about the behaviour or rationality of the experts involved. This contrasts with what occurs when reference is made to models based on preference relations and utility concept.

We assume alternatives independency and compute the consensus degree for each alternative as the weighted average of the linguistic performance values of each expert. The consensus level required for each alternative is fixed in advance as a percent of dispersion of the experts' linguistic performance values. If consensus has been reached for all the alternatives, the session for the consensus reaching can be closed. If this does not happen, the experts' opinions must be modified.

In our model the process of modifying opinions is managed by introducing the concept of opinion changing aversion (OCA). That is, we define for each expert a function representing his/her resistance to opinion changing. For expert k the function assumes the following form (see Fig.3):

$$f_k(x|V_k^*) = 1 - \frac{2}{e^{\frac{x-m_k}{a_k}} + e^{\frac{-x+m_k}{b_k}}}, \qquad (6)$$

where x assumes its values in the set M of the defuzzified values of experts' performances and m_k is the defuzzified value of V_k^*. The defuzzification method we use is the so called center-of-area method. The criteria that can be used to evaluate the best suitable method of defuzzification in this context are not taken into consideration here (see, e.g., [FILE91],[HELL93],[RUNK93]).

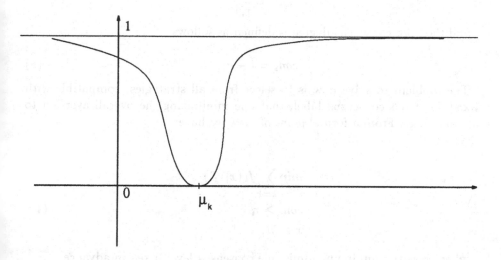

Fig. 3. function for opinion changing aversion

Furthermore we are not concerned here with how to set the shape of the OCA function, a problem that can be solved, e.g., on the basis of the Elementary Pragmatic Model [LEFO77],[SILV87] or according to information provided by the Source Control System [GARI88],[GARI92].

A similar approach was developed in classical decision analysis using the so-called state-dependent utility model (see, for instance, [DREZ87],[KARN87], [KELS92]).

Now the problem to be solved is to find, for each non-consensual alternative, a strategy which modifies the opinions of the experts in order to obtain a new set of individual performance values whose weighted mean is closer to the *consensual label*. The strategy chosen will be the t-ple $(V_{k_1}, V_{k_2}, \ldots, V_{k_K})$ that minimizes the overall aversion to opinion change. We define now the consensus degree, relative to alternative A_i, as a function both of the standard deviation (s_i) of the sequence of linguistic performance values of the experts and of the range (r_i) of the linguistic values V_l, over the universe of discourse V.

Formally we have

$$s_i = \left(\frac{1}{K-1} \sum_{k=1}^{K} [d(V_k^*, \mu_i)]^2 \right)^{1/2}, \tag{7}$$

where μ_i is the linguistic mean of the experts' performance values.
The range will be

$$r_i = \max_j\{\beta_j\} - \min_j\{\alpha_j\}, \tag{8}$$

normalizing we obtain

$$\sigma_i = \frac{s_i}{r_i},$$

and then the consensus degree is defined as follows

$$con_i = 1 - \sigma_i. \tag{9}$$

The problem to solve now is to select from all strategies, compatible with a fixed desidered consensus label, that one minimizing the overall aversion to opinion change. From a formal point of view we have:

$$min \sum_{k=1}^{K} f_k(x|V_l^*)$$
$$con_i \geq \tau_i \tag{10}$$
$$x \in M,$$

where τ_i is the minimum admissible consensus level, fixed in advance.

The moderator now invites the experts to shift their opinions towards the consensual label according to the optimal strategy $(\widehat{V}_{k_1}, \widehat{V}_{k_2}, \ldots, \widehat{V}_{k_K})$. After the experts have changed their opinions, the new degree of consensus is computed and if it is greater than or equal to the minimum admissible level the process stops. Otherwise the parameters of the program (10) are updated and a new solution is looked for. The interactive and iterative process will finish when for each alternative the minimum level of consensus or a deadlock is reached.

4.2 Simulation

An experimental study of the OCA-based approach was carried out in order to establish its performance. Since the results of realistic study cannot be generalized, we organize a set of simulated consensus-reaching sessions.

We developed a computer program to analyze the behaviour of the algorithm we intended to propose. The simulation used randomly generated data and the evaluations of the experts were modified by an *ad-hoc* internal routine. The algorithm was evaluated using contingency tables comparing two approaches, the one of the model and a random one.

The preliminary results showed that the model-based approach obtained better results in most cases.

The simulation was performed on an MS-DOS computer with an i486 CPU, 25 Mhz, 4 Mbyte RAM, 120 Mbyte using the following tools: MsDos 5.0, Microsoft Fortran 5.1, Imsl Math & Stat Libraries. The module for the management of conflicts and a GDSS prototype was implemented in a Windows 3.1 environment using Visual Basic 1.0 as development tool.

References

[BOAT91] Boatto, V., Fedrizzi, M., Furlani, P., Povellato, A.: Fuzzy logic in group decisions: an application to problems of environmental policies. Land and Resources Use Policy (1991) 62-82

[BONI82] Bonissone, P.P.: A fuzzy sets linguistic approach: theory and applications. In: Gupta, M.M., Sanchez, E. (Eds.) Approximate Reasoning in Decision Analysis. North-Holland (1982) 329-339

[BORT85] Bortolan, G., Degani, R.: A review of some methods for ranking fuzzy subsets. Fuzzy Sets and Systems. North-Holland 1 (1985) 1-19

[BUI 87] Bui, T.X.: A Group Decision Support System for Cooperative Multicriteria Group Decision Making. Lecture Notes in Computer Science. Springer-Verlag 290 (1987)

[CARL92] Carlsson, C., Ehrenberg, D., Eklund, P., Fedrizzi, M., et al.: Consensus in distributed soft environments. European Journal of Operational Research 61 (1992) 165-185

[DESA87] Desanctis, G., Gallupe, B.: A foundation for the study of group decision support systems. Management Science 33 (1987) 589-609

[DREZ87] Dreze, J.M.: Essays in Economic Decisions under Uncertainty. Cambridge University Press (1987)

[DUBO82] Dubois, D., Prade, H. The use of fuzzy numbers in decision analysis. In: Gupta, M.M., Sanchez, E. (Eds.) Fuzzy Information and Decision Processes. North-Holland (1982) 309-321

[DUBO87] Dubois, D., Prade, H.: Fuzzy numbers: an overview. In: Bezdek, J.C. (Ed.) Analysis of Fuzzy Information. CRC Press (1987) 3-39

[EOM 90] Eom, H.B., Lee, S.M., Suh, E.-H.: Group decision support systems: an essential tool for resolving organizational conflicts. International Journal of Information Management 10 (1990) 215-227

[FEDR88] Fedrizzi, M., Kacprzyk, J., Zadrozny, S.: An interactive multiuser decision support system for consensus reaching processes using fuzzy logic with linguistic quantifiers. Decision Support Systems 4 (1988) 313-327

[FEDR91] Fedrizzi, M., Mich, L.: Decision using production rules. In: Proc. of Annual Conference of the Operational Research Society of Italy (AIRO'91). September 18-10. Riva del Garda. Italy (1991) 118-121

[FEDR92] Fedrizzi, M., Fuller, R.: Stability under fuzzy production rules. In: Proc. of 11th European Meeting on Cybernetics and Systems Research (EMCSR'92). April 21-24. Vienna. Austria (1992)

[FILE91] Filev, D.P., Yager, R.R.: A generalized defuzzification method via bad distributions. International Journal of Intelligent Systems 6 (1991) 687-697

[GARI88] Garigliano, R., Long, D., Bokma, A.: A model for learning by source control. In: Bouchon, Saitta and Yager (Eds.) Uncertainty and Intelligent Systems. LNCS 313. Springer-Verlag (1988)

[GARI92] Garigliano, R., Bokma, A.: Heuristics for knowledge extraction in Source Control. International Conference on Information Processing and Management of Uncertainty (IPMU'92). July 6-10. Mallorca. Spain (1992) To be published in a LNCS volume

[GRAY87] Gray, P.: Group decision support systems. Decision Support Systems 3 (1987) 233-242

[HELL93] Hellendoorn, H.: Design and development of fuzzy systems at Siemens R&D. Second IEEE International Conference on Fuzzy Systems. March 28 - April 1. San Francisco. USA (1993) 1365–1370

[HUBE84] Huber, G.P.: Issues in the design of group decision support systems. MIS Quarterly 8 (1984) 195-204

[JACO92] Jacob, V.S., Pirkul, H.: A framework for supporting distributed group decision-making. Decision Support Systems 8 (1992) 17-28

[KACP88] Kacprzyk, J., Fedrizzi, M.: A 'soft' measure of consensus in the setting of partial (fuzzy) preferences. European Journal of Operational Research 34 (1988) 316-325

[KARN87] Karni, E.: Generalized expected utility analysis of risk aversion with state-dependent performances. International Economic Review 28 (1987) 229-240

[KELS92] Kelsey, D.: Risk and risk aversion for state-dependent utility. Theory and Decision 33 (1992) 71-82

[KOWI80] Kowitz, A.C., Knutson, T.J.: Decision Making in Small Groups. Allyn and Bacon. Boston (1980)

[KRAE88] Krämer, K.L., King, J.L.: Computer based systems for cooperative work and group decision making. ACM Computing Surveys 20 (1988) 115-146

[LEE 83] Lee, N.: Environmental impact assessment: a review. Applied Geography 3 (1983)

[LEFO77] Lefons, E., Pazienza, M.T., Silvestri, A., Tangorra, F., Corfiati, L., De Giacomo, P.: An algebraic model for systems of psychically interacting subjects. In: Dubuisson, B. (Ed.) Information and Systems. Oxfor. Pergamon Press (1977) 155-163

[LEVI74] Levi, I.: On indeterminate probabilities. Journal of philosophy 71 (1974) 391-418

[LOEW85] Löwer, B. (Guest ed.): Special Issue on Consensus. Synthese 62 (1985)

[MART86] Martel, J.M., D'Avignon, G.R., Coillard, J.: A fuzzy outranking relation in multicriteria decision making. European Journal of Operational Research 25 (1986) 258-271

[NUNA91] Nunamaker, J.F., Dennis, A.R., Valachic, J.S., Vogel, D.R., George, J.F.: Electronic meeting systems to support group work. Communications of the ACM 34 (1991) 40-61

[ROY 77] Roy, B.: Partial preference analysis and decision aid: the fuzzy outraning relation concept. In: Bell, D., Keeney, R., Raiffa, H. (Eds.) Conflicting objectives in decision. AC-Press (1977)

[RUNK93] Runkler, T.A., Glesner, M.: A set of axioms for defuzzification strategies - towards a theory of rational defuzzification operators. Second IEEE International Conference on Fuzzy Systems. March 28 - April 1. San Francisco. USA (1993) 1161–1166

[SAAD92] Saade, J.J., Schwarzlauder, H.: Ordering fuzzy sets over the real line: an approach Bbased on decision making under uncertainty. Fuzzy Sets and Systems 50 (1992) 237-246

[SHAK91] Shakun, M.: Airline buyout: evolutionary system design and problem restructurinh in group decision and negotiation. Management Science 37 (1991) 1291-1303

[SILV87] Silvestri, A., Mich, L., Pereira Gouveia, O., Ferreira Pinto, C.: Simulation of the development of individual interactional patterns. Cybernetics and Systems: An International Journal 18 (1987) 489-515

[TAKA92] Takahara, Y., Iijima, J., Shiba, N.: A hyerarchy of decision making concepts: conceptual foundation of decision support systems. International Journal of General Systems 20 (1992) 359-378

[TONG80] Tong, R.M., Bonissone, P.P.: A linguistic approach to decision making with fuzzy sets. IEEE Transactions on Systems Man and Cybernetics 10 (1980) 716-723

[ZADE73] Zadeh, L.: Outline of a new approach to the analysis of complex systems and decision processes. IEEE Transactions on Systems Man and Cybernetics 3 (1973) 28-44

[ZADE75] Zadeh, L.: The concept of a linguistic variable and its application to approximate reasoning. Information Sciences (Part I) 8 (1975) 199-249 (Part II) 8 301-357 (Part III) 9 (1976) 43-80

[ZIMM91] Zimmermann, H.J.: Fuzzy Set Theory and its Applications, 2nd Ed., Kluwer, Dordrecht-Boston, 1991

Fuzzy Logic-Based Processing of Expert Rules Used for Checking the Creditability of Small Business Firms

HEINRICH J. ROMMELFANGER

J.W.Goethe-University Frankfurt am Main
Department of Economics and Business Administration
Institut of Statistics and Mathematics
Mertonstraße 17-25, D-6000 Frankfurt am Main 11

Abstract. In this paper is analysed, whether and how the proceeding commonly usued in Fuzzy Control can be transformed to non-technical expert systems. The new ideas are explained by means of a system for checking the creditability of small business firms.

1 Introduction

Having in mind, that fuzzy controllers proved to be superior to traditional methods in many technical applications and that in Japan the term "fuzzy" stands for progressive technique, the question arises whether fuzzy control can also be applied to non-technical expert systems. Undoubtedly only few fuzzy control programmes can strictly speaking be considered expert systems. Especially self-organizing controllers in camcorders, telescopes, cameras etc. cannot be classified as expert systems. Other applications, however, as train control or the optimizing control of a Diesel engine, doubtless are expert systems because they satisfy more than one of the following four criterias which according to Schnupp/Nguyen Huu [10] distinguish expert systems from traditional systems:

- the system possesses a knowledge base which means it accumulates not only facts (data base), but also rules (production rules),
- the system contains components which cultivate and extend the knowledge base (passive and active capability of learning),
- basing on the facts and rules stored in the knowledge base the system can produce new knowledge by using heuristics (processing techniques and search strategies),
- the system is able to explain the procedure chosen to solve a problem as well as name arguments in favour of the presented solution (explanatory component).

In technical control systems only the rule map contains unprecise information in general. This inaccuracy is usually caused by the expert who either does not know all the relevant cognitive processes, does not mention them at knowledge acquisition or does simply not express his knowledge explicitly enough. In expert systems for decision support of non-technical problems more uncertainties frequently become evident.

The following can be named:
- decision variables are unknown or are neglected due to the complexity of the problem,
- the rule map often is incomplete in consequence of missing information (in comparison to that rules are consciously omitted in fuzzy control models in order to increase the control speed),
- data-basis is inaccurate, vague data,
- information is not sufficient for specifying probabilities.

Considering the abundance of unprecise information it is clear that very early attempts were made to integrate fuzzy concepts into expert systems. Zimmermann [13] quotes 24 expert systems by name which make use of fuzzy sets in various forms. During the past years additional systems took advantage of this technique. Fields of application are pattern recognition, medical diagnosis, treatment of Diabetes, chest pain diagnosis, mineral exploration, business planning, creditability check and many more.

Most of the fuzzy expert systems make use of several fuzzy concepts. Among the well known ones are:
- the use of linguistic variables in order to describe the terms of attributes (i.e. good, medium, poor), frequencies (i.e. never, rarely, normal, frequent, always), the actions (i.e. strong, medium, weak), truth qualification (i.e. true, false),
- the application of fuzzy numbers and fuzzy intervals in order to describe vague data and their conjunctions by means of extended operators, for example to define fuzzy expectation values or fuzzy expectation intervalls,
- the use of fuzzy logic concepts which replace the clear-cut terms "true" and "false" by corresponding linguistic variables in the sense of Zadeh and Baldwin,
- the employment of the possibility theory if there exists no sufficient information to define probabilities,
- the application of membership grades in order to evaluate objects in a hierarchic system, and the use of fuzzy operators to aggregate these valuations.

2 Hierachical Aggregation of Determinants of Creditworthiness

The concept last mentioned was used in an empirical research project of the Institut of Statistics and Mathematics, University of Frankfurt/Main in 1985- 1987. The aim of this research work was to explain how credit manager of banks arrive at a decision concerning the creditworthiness of small business firms. Based on a first empirical study a hierarchical systems of determinants of the material creditworthiness was constructed, see Rommelfanger/Unterharnscheidt [6,.7] and the newer hierarchy of 1991 in Figure 1. By a second empirical study, the relative importance of the determinants on the different levels was investigated. On the basis of a third extensive study at which 50 credit experts evaluated the creditworthiness of 30 firms it should be explained, how credit manager aggregate their judgements concerning the determinants of creditworthiness for arriving at a creditability decision.

Assuming that the human evaluations of the determinants of creditworthiness can be represented by membership degrees the formal connectives of fuzzy sets can be con

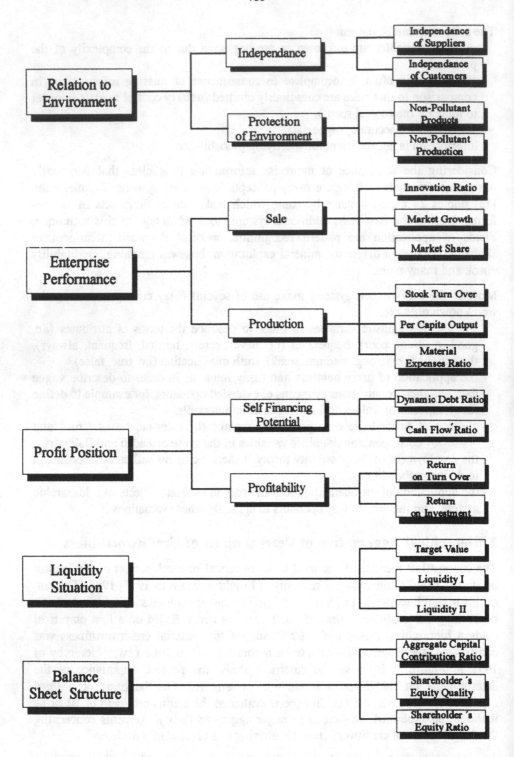

Fig. 1: Hierarchical System for Evaluating Material Business Creditability

siderd as models for the aggregation process. 14 fuzzy operators (without and with weights) were tested with the result, that no one of these aggregation operators had a sufficient good predictive quality **in all parts** of the hierachical system. One of the best was the simple arithmetic mean which satisfied the demanded predictive quality in 26 of 28 cases, see Rommelfanger/Unterharnscheidt [7]. One explanation for this result is that the inquired credit managers used this simple procedure in view of the complex decision situation.

By analysing the data in detail we recognized that not only the determinants but also their evaluations (the membership degrees) influenced the aggregation process. And this result is not new, in literature and practise there exist a lot of examples, that the weigts in an objective system depend not only on the objectives but often change with the obtained values. Therefore a so called asymmetric operator was developed 1991 which offers the possibility that the weights of the particular determinants may depend on the obtained membership grade. This character qualifies them for strategic planning models and for models which intend to evaluate capital structure, financial assets and revenue of firms, see Paysen [4] and Scheffel [8]. But referring to the empirical study "creditability of small bissiness firms" the asymmetric operater yielded to results which were hardly better than these of the arithmetic mean.
All in all it became apparent that the application of parameter dependant operators reflect the complex conjunction mechanism of the human mind only incompletely. Therefore it is necessary to look for other ways for modelling the decision process of credit managers.

3 Aggregation of Determinants by Expert Rules

In artificial intelligence-literature we can find many models where the human decision process is described by means of rules, which are formulated by experts. The following tables 1 and 4 present two rule maps stated by an expert team of the Commerzbank AG Frankfurt am Main. These tables are part of a new developed model for creditability checks. The Figure 1 shows the hierarchical system of the part which shall be used to evaluate the material business creditability, see Bagus [1].

Table 1. Aggregation Rules of Self Financing Potential

CF-Rate	Dyn. DR	Self Fin. Pot.
p	p	p-
p	m	p
p	g	m-
m	p	p
m	m	m
m	g	m
g	p	m
g	m	g
g	g	g+

Table 2.	Valuation of Cash Flow-Rate		Table 3.	Valuation of Dynamic Debt Ratio	

CF-Rate	Grade
< 0%	(6)
0% - 2%	6
2% - 4%	5
4% - 6%	4
6% - 8%	3
8% - 10%	2
> 10%	1

Dyn. DR (Years)	Grade	
> 10	6	poor
8 - 10	5	(great risk)
6 - 8	4	medium
4 - 6	3	(medium risk)
2 - 4	2	good
< 2	1	(small risk)

In this rule-based model the evaluation of determinants is not done by membership degrees of the interval [0, 1] but by the linguistic terms "poor", "medium" and "good". For every possible situation on the lower hierarchy level an aggregation rule was formulated. In this study only three values per criterion were distinguished, however additional ratings (- and/or +) were allowed for the aggregation results.

Objections against those rules can be raised because the rules are very inaccurate,the terms "good", "medium", "poor" offer a comparably large interpretational spectrum, see the Tables 2, 3, 5 and 6. To demonstrate this disadvantage we consider the following three firms:
Firm A : Cash Flow-rate 4.1%, Dynamic Debt Ratio 7.9 years
Firm B : Cash Flow-rate 7.9%, Dynamic Debt Ratio 4.1 years
Firm C : Cash Flow-rate 3.9%, Dynamic Debt Ratio 4.1 years
According to the rule map in Table 1 the firms A and B get the same evaluation "medium" for the determinant "Self Financial Potential", though the firm B is much better in both basic criteria. On the other side, the firm C gets the worse valuation "poor", though it has a better valuation for the critirion "Dynamic Debt Ratio" than firm B and the valuation of "Cash Flow-Rate" is only insignificant worse. Moreover the strong distinctions between the classes give the impression of an arbitrary classification.

To improve this situation we could try to enlarge the number of evaluations for each criterion. But this would have the consequence that the number of rules will increase explosively because for m aspects with r possible evaluations there exist m^r rules. Therefore it is necessary to limit the number of valuations for each aspect. Moreover if the rule map gets too large it will not be possible for the expert team to guarantee a conscious distinction of each situation.

Table 4. Valuation of State of Liquidity

Liqu. I	Liqu. II	Target Value	State of Liquidity
p	p	p	p-
p	p	m	p-
p	p	g	p
p	m	p	p
p	m	m	m
p	m	g	m+
p	g	p	m
p	g	m	m+
p	g	g	g-
m	p	p	p
m	p	m	p
m	p	g	m-
m	m	p	p
m	m	m	m+
m	m	g	m+
m	g	p	m
m	g	m	g
m	g	g	g
g	p	p	p+
g	p	m	m
g	p	g	m+
g	m	p	m
g	m	m	g
g	m	g	g+
g	g	p	g-
g	g	m	g+
g	g	g	g+

Table 5: Valuation of Liquidity

Percentage of Deviation	Grade
< 50%	6
50% - 60%	5
60% - 70%	4
70% - 80%	3
80% - 100%	2
> 100%	1

Table 6: Valuation of Target Value

Percentage of Difference	Grade
> 30%	6
15% - 30%	5
0% - 15%	4
-15% - 0%	3
-30% - -15%	2
< -30%	1

4 Description of Expert Rules by Fuzzy Sets

As demonstrated in the example above different values of the same interval express a linguistic term not to the same extent. This is conditioned by the fact that intervals only allow Yes/No-statements. Therefore we propose to model the linguistic terms by fuzzy sets; this theory makes it possible to describe the different membership degrees according to the categories of "poor", "medium" or "good" creditworthiness as precise as the credit expert may express it. These membership functions must be specified as carefully as possible, because this can decisively influence the evaluation process. Nevertheless we can not expect membership functions which are precise in every detail, because a lot of data about similar firms and knowledge of the trade must be composed by the expert team. Therefore the form of membership functions in expert systems will be very simple and the same design will used repeatedly. In practice it is sufficient to work with fuzzy numbers or fuzzy intervals of the RL-type, see [5]. Whereas in Fuzzy Control applications the very simple triangular or trapezoid fuzzy sets are mainly used, we propose to use reference functions of the s-type, which take pattern to normal probability distribution, in non-technical applications. In the Figures 2, 3, 4 the reference function $\exp(-u^2)$ is used. To specify the form of a membership function for small membership degrees is particulerly difficult. To avoid errors we propose to neglect all membership degrees lower than a minimum level ε, in the Figures 2 - 4 this minimum level is fixed as $\varepsilon = 0.05$.

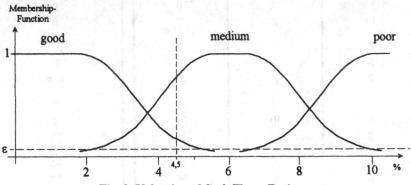

Fig. 2. Valuation of Cash Flow - Ratio

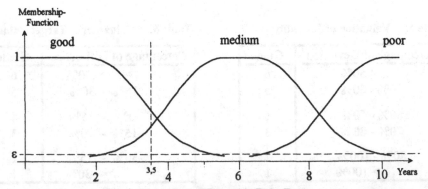

Fig. 3. Valuation of Dynamic Debt Ratio

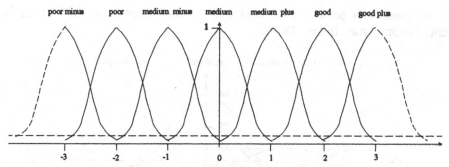

Fig. 4. Valuation of Self Financing Potential

5 Fuzzy Controller and Rule-Based Aggregation

Modelling the linguistic valuation terms by fuzzy sets the aggregation rules are applied only to the cases, in which for all subaspects one of the valuations is fulfilled with the membership degree 1. Then the corresponding rule in the rule map is used and it leads to an unique valuation of the upper-aspect with the membership degree 1 too. Therefore using fuzzy sets enables users to understand the basic principles of the expert knowledge more easily. This "understanding" is an essential factor for the acceptance of an expert system and related with that its successful realization.

As an example, we contemplate a firm with the characteristics

(Cash Flow Rate , Dynamic Debt Ratio) = (11% , 6 Years).

Then according to the 7th rule in Table 1 the Self Financial Potential get the valuation "good" with the membership degree 1.

For all the other cases, where at least one valuation has a membership degree smaller than 1, no special rules are stated by the experts. But we assume that the given rules can be extended to situations in the neighbourhood. The rules are softene with the consequence that now many rules can be used simultaneously in a weakened manner.

For a real situation we denote the degree of fulfilment with the descriptions of state in the rule maps by DOF. According to the proceeding in fuzzy control, DOF is defined as the minimum of the membership degrees attached to the "inputs" of this rule. Though we examined whether other operators would better describe the human conjunction behaviour in specific cases we came to the result that the minimum operator should used. E. g. it has the advantage that only few rules with positve DOFs exist, whereas by using compensatory operators almost all rules will show positive DOFs and therefore we have to expect intermediate valuation.

For example, a firm D with the characteristics

(Cash Flow Rate , Dynamic Debt Ratio) = (4.5% , 3.5 Years)

get according to Table 1 the following four positive DOFs:

$DOF_{Regel\ 2} = Min\ (\mu_{poor}(4.5)\ ,\ \mu_{med}(3.5)) = Min(0,13\ ,\ 0.27) = 0,13$ *poor*

$DOF_{Regel\ 3} = Min\ (\mu_{poor}(4.5)\ ,\ \mu_{good}(3.5)) = Min(0.13\ ,\ 0.48) = 0,13$ *med -*

$DOF_{Regel\ 5} = Min\ (\mu_{med}(4.5)\ ,\ \mu_{med}(3.5))\ \ = Min(0.72\ ,\ 0.27) = 0.27$ *med*

$DOF_{Regel\ 6} = Min\ (\mu_{med}(4.5)\ ,\ \mu_{good}(3.5)) = Min(0.72\ ,\ 0.48) = 0.48.$ *med*

Now all rules with positive DOF contribute to the valuation of Self Financial Potential in proportion to their DOFs.

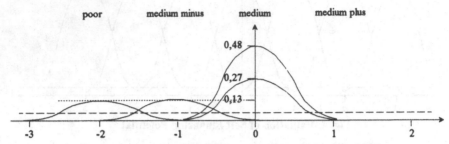

Fig. 5. Valuation of Self Financial Potential using Max-Prod-Inference

In "turning down" the valuations corresponding to the individual rules we recommend the use of the Max-Prod-inference which means that the membership values are fixed in proportion to the corresponding DOF, see figure 5. We are convinced that the Max-Prod-inference is better than the Max-Min-inference which is often used in Fuzzy Control applications, because the elimination of membership values which go above the DOF implies that rules with medium DOF get more influence.

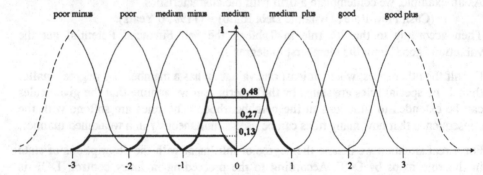

Fig. 6. Valuation of Self Financial Potential using Max-Min-Inference

In this context we want to remark that there is an essential difference between fuzzy control applications and non-technical valuation and decision problems. Usually technical control processes are rapidly repeated, see [11, 12]. Therefore it is sufficient when an approximately correct action is carried out, because the correction will follow immediately. Decision support systems require a definite decision for every section which evidently has to be correct. As a consequence not only the linguistic valuation terms have to be defined more carefully but also the calculation of the DOFs and the influence of the DOFs on the final result need an exact empirical examination.

The total valuation - by application of the Fuzzy Control procedure - results from the fusion of the valuations of the relevant rules by using the Maximum-operator. But in decision-support-systems corrections should be considered. In our example the valuation "medium" for the Self Financing Potential can be found twice. On the one hand we do not consider it to be right that the rating "medium" only counts with a DOF Max(0,27, 0,48) = 0,48 which means that the rule with the second best positive DOF is completely neglected. On the other hand it seems absurd to add the DOFs if two or more rules turn up with the same "output", because it would then be possible to get DOF-values greater then 1. We propose to adopt a middlle course and suggest the use of the algebraic sum. In doing so in the example above, we get for the valuation "medium" the total DOF 0,27 + 0,48 - 0,27 * 0,48 = 0,62 which presents a more balanced valuation.

Then the self financing potential is vaguely evaluated by means of the membership function shown in figure 7.

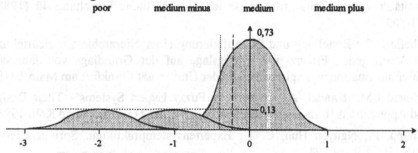

Fig. 7. Valuation of Self Financing Potential for the firm D

If it is intended to "compress" the data to a unique valuation, the center of gravity method can be applied to calculate a "mean" value, it is marked by little lines in Figure 7. Another choice presents the plane bisection method which offers in our example a slightly higher result, which is marked by the pointed line in Figure 7. But in hierachical systems defuzzifying is not necessary, because it is better to use directly the fuzzy valuations as inputs for the next aggregation step.

6 Final Remarks

The application of linguistic variables and the employment of fuzzy conjunction methods offer an appropriate method to structure the human reflection process and by doing so to construct expert systems which actually deserve this name. There are still a lot of questions left to be answered, but I am convinced that this concept aims at the right direction and that the first applications will be used in practice soon.

7 References

1. Bagus, Th.: Wissensbasierte Bonitätsanalyse für Firmenkundengeschäft der Kreditinstitute. Peter Lang-Verlag Frankfurt a. M. 1992
2. Braun, F.: Fuzzy-Verknüpfungsoperatoren mit asymetrischer Kompensation - Theorie und empirische Überprüfung. Diplomarbeit an der Universität Frankfurt am Main 1991

3. Hall, L. O., Kandel, A.: Designing Fuzzy Expert Systems. Verlag TÜV Rhein Köln 1986

4. Paysen, N.: Unternehmensplanung bei vagen Daten. Peter Lang-Verlag Frankfurt am Main 1992

5. Rommelfanger, H.: Entscheiden bei Unschärfe.- Fuzzy Decision Support-Systeme. Springer-Verlag Berlin Heidelberg 1988

6. Rommelfanger, H., Unterharnscheidt, D.: Entwicklung einer Hierarchie gewichteter Bonitätskriterien für mittelständische Unternehmen. Österreichisches Bank-Archiv 33 (1986), 419-437

7. Rommelfanger, H., Unterharnscheidt, D.: Modelle zur Aggregation von Bonitätskriterien. Zeitschrift für betriebswirtschaftliche Forschung 40 (1988), 471-503

8. Scheffels, R.: Erstellung und Quantifizierung einer Hierarchie zur Beurteilung der Vermögens-, Finanz- und Ertragslage auf der Grundlage von Jahresabschlußinformationen, Diplomarbeit an der Universität Frankfurt am Main 1991

9. Schneider,M., Kandel, A.: Cooperative Fuzzy Expert Systems - Their Design and Applications in Intelligent Recognition. Verlag TÜV Rheinland Köln 1988

10. Schupp, P., Nguyen Huu, C. T.: Expertensystempraktikum. Springer-Verlag Berlin Heidelberg 1987

11. Sugeno, M. (Ed.): Industrial Applications of Fuzzy Control. North Holland Amsterdam 1985

12. Yamakawa, T.: Stabilization of an Inverted Pendulum by a high-speed Fuzzy Logic Controller Hardware System. Fuzzy Sets and Systems 32 (1989), 161-180

13. Zimmermann, H. -J.: Fuzzy Sets, Decision Making and Expert Systems. Kluwer Academic Publishers, Boston 1987

Fuzzy Control in Real-time for Vision Guided Autonomous Mobile Robots

B. Blöchl[*]

Ruppertstr. 8, 8000 Munich 2,
Germany

Abstract. To refine lateral control in an autonomous vehicle, a number of fuzzy controllers were developed in real-time simulation trials using real-world video images. The controller with the best results from the simulation trials was then mounted on the experimental vehicle ATHENE and shown to be functionally sound under the real-world circumstances of letting it drive through a corridor. The results were recorded and evaluated. One particular advantage of the fuzzy controller became obvious: input of the required quality could be derived relatively easily from the video images by means of simple image processing without extensive camera calibration. The development environment for the simulation and image processing are described in this paper, and the results are presented and explained.

1 Introduction

The human being relies primarily on his sight for the successful execution of almost all activities. During the course of evolution, this capacity has been highly developed for humans as well as for most other animal forms. Robots possessing the capacity for sight will be those able to attain the highest degree of autonomy. A prerequisite for the introduction of autonomous robots into real-world activities, however, is their capacity to visually process a situation and evaluate it in real time. Specifically, this involves object recognition, assessment of the information in light of the situation itself, and then the ability to select a reasonable and appropriate reaction. Of the tasks involving the image processing system for an autonomous vehicle, the most important are the recognition and continuous trakking of paths in front of the vehicle. These functions are part of a fundamental feedback mechanism that keeps the vehicle on the path.

To implement a course-holding function using a conventional control algorithm, the relevant information must be extracted from raw image data of the course, then prepared as input for the algorithm. A central problem in this process is the high expenditure involved in the camera calibration, which is necessary in order to produce a distinct geometric correspondence between the environment and the camera image. A conventional controller requires rigidly exact input, an exact system description, and an unambiguous arithmetic correlation. The calculations involved require intensive processing power. A second problem is that the system

---

[*]former Universität der Bundeswehr, Institut für Meßtechnik, 8014-Neubiberg, Germany

dynamics change each time a vehicle is even slightly modified, such as when a lorry and carrier is loaded, a typical occurrence. The system dynamics defined by the control algorithm is not flexible enough to be able to compensate for such abrupt changes to a vehicle.

The human being, on the other hand, reacts to a situation using information he gathers on the basis of quality rather than quantity. His perception of the system itself is also based on a qualitative assessment. Any changes to the system are perceived and taken into consideration in the reaction. The human's control strategy consists of simple rules brought to bear on the controlling action manually. In order to keep to a lane, a human driver turns the steering wheel to the right if the vehicle veers too far to the left. And in the opposite case, he steers to the left if the vehicle veers too far to the right. Steering a vehicle in this way (left and right) defines our entry variable. The corresponding exit variable is the steering angle. The human determines the degree of course correction (the control variable) intuitively, depending on how far from the lane the vehicle has moved and how fast the vehicle is moving. This is the "expert knowledge " to be modelled for the autonomous vehicle. For the human, it is embodied in the basic everyday *If-Then* rule. This rule is applied to the control variable, what the human understands as the necessary degree of course correction, and what is referred to in this paper simply as the "steering angle".

This rule allows for the exploitation of inference rules, and after the consequent defuzzification, the positioning element for the vehicle receives the output: the size of the steering angle to be executed. Although a conventional controller can be used to implement this process only with difficulty, fuzzy logic is suited for such tasks. A fuzzy controller exploits a technically adapted application of fuzzy logic.

As the results of this study show, fuzzy control will steer a vehicle to its goal in situations posing impossible constraints both for system models and for conventional control techniques which are brought to their limits: situations requiring nonlinear, temporally variable behavior yet nevertheless providing measurements of only low quality.

2 System Structure

In a vehicle's immediate surroundings, the relevant objects must be not only recognized but also perceived in relation to each other. The vehicle must be able to react appropriately depending on the situation.

To understand any one scenario, several tasks must be performed: all relevant objects must be recognized and described or prepared for the process of description; a general description must be produced from a meaningful combination of object descriptions, and this general description must consistently result in an accurate evaluation of the scenario. From this situation evaluation, an appropriate action must be executed, which is then delivered to the actuator by means of a controller.

This paper describes how path tracking is performed by fuzzy control. A full description of system requirements for intelligent autonomous mobility, however, is not provided. A compact overview of the extended system aspects for an intelligent automatic copilot to support human drivers on motor ways with multiple lanes is given in [Graefe, Blöchl 1991]. For a description of the hardware requirements for a comparably intelligent system, refer to [Graefe 1984; Graefe, Kuhnert 1992; Blöchl 1992].

Unusually high demands are placed on the computing powers of a system for an autonomous mobile robot as far as image processing, controller and decision-making (intelligent basis) are concerned. In real time, the work cannot be accomplished with a single pro-

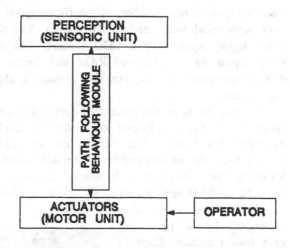

Fig. 1 Simplified system with a single behaviour module for path following.

cessor. A division of the work is unavoidable. If tradition were to be followed, the problem would be approached by subdividing the sum of work into a number of functional units. [Brooks 1986] poses an alternative concept (Fig. 1), the subdivision of work into what he calls task-achieving behaviors.

3 Path Finder Module

A very simple concept emerges when one considers the straightforward goal of following a path without consideration of obstacles. ("Consideration of obstacles" would require two modules: one module to recognize obstacles; a second module either to alter the vehicle's course to avoid the obstacle or to halt the vehicle.) The simpler concept is made up of a single behavior module (Fig. 1) consisting of two cooperating submodules:

▶ an image-processing (recognition) module to extract information of the path from the raw pixel data provided by a CCD camera.

▶ a fuzzy-control module to control the actuators of the vehicle in order to achieve the desired behavior.

In autonomous driving on a highway or elsewhere, the lane or path being used is the most essential object to be recognized, both for human drivers and, as in this case, for the autonomous vehicle. The lane or path is to be determined by means of image processing. The video image is supplied by a CCD camera mounted on the front of the vehicle (see Fig. 2 and 3.). For the minimized task of path tracking, a single Transputer-framegrabber (TFG-Ultra of Parsytec) equipped with a single transputer (T805) is adequate. The computing performance of the single transputer is sufficient to process the images and to operate the fuzzy controller.

The Transputer-Framegrabber, originally designed as a card for a PC, makes for a very economical and user-friendly system. The card digitizes the standard PAL video signal supplied by a CCD camera. The transputer receives the digitized video signal via a dual-ported RAM and during image processing ascertains the driving lane from the raw pixel data stream. It also performs the fuzzy control of the vehicle.

To track the path, the controlled-correlation method was used. This is a generalization of the well-known correlation method [Kuhnert 1988; Graefe, Tsinas 1992, Wershofen 1992] suitable for real-time applications. The method has already been applied successfully to the autonomous tracking of a lane in a road at 96 km/h, this limit being set by the vehicle's motor, rather than by any constraint imposed by the image processing [Dickmanns, Graefe 1988; Zapp 1988].

Skirting boards are characterized by a prolonged grey tone transition (light to dark) in the picture, with angles of 45 degrees (left) and 135 degrees (right). Customized feature operators (bit templates) were used to search for these edges. The position of the edges in the picture served as input to the fuzzy controller. The controller's task is to maintain these edges at a suitable position in the picture. The experiments described here as well as those referenced above were performed without the use of camera calibration, proving the robustness of the methods developed.

The wide-angle lens (8 mm) used with the CCD

Fig. 2 A corridor at the laboratory site as seen from the vehicles CCD camera. The skirting boards compare to the painted lane markers on a motorway.

sensor, reaches a horizontal visual angle of 58 degrees. The sensor's horizontal resolution of 683 pixels results in an angle of 0.085 degree/pixel. With this resolution, the angle of deviation between the centre of the camera image (optical axis) and the centre of the road (vanishing point) can be determined. Stable behavior was also achieved when the controller was fed with a value ten times higher than the correct one, thus demonstrating the robustness of the fuzzy controller.

4 The Development Environment

Real-time simulation proved worthwhile for the development of both an image-processing algorithm as well as for a controller. In this real-time simulation, the framegrabber was provided with images of an experimental drive from a video recorder. The succession of pictures, taken by a conventional black-white CCD camera, were exposures of a length of the experimental track. The advantage is obvious: a scenario can be repeated as often as desired, free of the unexpected occurrences typical of real-world experiments. All analysis and error identification are thereby significantly simplified. For the development of image processing algorithms, the image sequence as well as the images from the video record are provided and processed in real time, thereby producing an equivalent to a real-world trail run.

To develop and test the controllers in this simulation environment, a slightly modified procedure is necessary requiring a vehicle dynamic. In the simulation, it was assumed that a portion of the video image corresponded to the image produced by the camera mounted on the vehicle. A 156 x 156 pixel "cut out" from a total image the size of 256 x 256 pixels proved to be effective. The movement of the vehicle is simulated by shifting the cutout over the surface of the entire image. Using these image proportions and a focal length of 25 mm, it was possible to simulate the vehicle's movement up to ± 4 ° (Larger angles would allow the cutout to stray from the basic 256 x 256 picture.) The movement of the cutout corresponds to the movement of the vehicle with its mounted camera. For the dynamic vehicle model, a straightforward correlation between position and degree of course correction is assumed. This simple linear model is often used, but best describes the behavior of a vehicle executing small steering angles. (The degree of course correction in the simulation was limited to ± 4 ° for this reason.) System dynamics of much greater complexity can also be modelled, so long as the limits imposed on the angle causes no difficulty. For our experiments, the linear model was sufficient.

With that our simulation environment was complete, for the development of image processing and the controller, and for the experimental vehicle trials. For this simulation environment, the performance capacity of the above described Transputer-framegrabber was adequate. The image processing and the controller work together just as they would in a real vehicle. In the simulation, the controller shifted the cutout in response to necessary course corrections, whereas in real systems the vehicle's actual course would be changed.

5 The Experimental Vehicle

For safety reasons, only tested and approved concepts may be applied in normal public road traffic. To enable the efficient development and safe testing of new basic concepts, a real-time simulation with video scenes was carried out in the first step. The controller developed as a result of this first step was then tested in indoor experiments for the vehicle ATHENE. This vehicle made laboratory experiments possible under real-world, real-time conditions, including unpredictable disturbances.

The experimental platform ATHENE (Fig. 3) is a three-wheeled vehicle, approx. 1.5 m long and 0.7 m wide, with front-wheel drive. The maximum velocity is approx. 1.5 m/s. Each rear wheel is provided with an odometer. On the front and sides there are bumpers for emergency stops in the case of a collision. The main sensor is a CCD camera mounted on a platform on the vehicle's front. For controlling the actuators, the platform is provided with a separate system computer, based on an INTEL 80188 microcontroller. The system computer has a cycle time of 110 to 170 ms. The cycle time of image processing is 20 ms. The cycle time of the experimental platform, however, determines the transmission rate of control data.

Fig. 3 A sideview of the experimental vehicle ATHENE used in the experiments with the control monitors on top and a CCD camera at the front of the vehicle.

Compared to the inert 5-ton lorry also available for experiments at the university site, the faster ATHENE in combination with the long cycle time poses a rigorous test for control algorithms. Experiments with the lorry require a great deal of preparation, a large crew, and are time-consuming. More efficient tests can be performed with the ATHENE.

ATHENE requires a restricted steering angle to achieve results that can be compared to those of heavy vehicles. From reports dealing with autonomous robots [Nishiki-cho, Chiyoda-ku, 1991] and as a result of experiments with vehicles, a change rate of the steering angle of 8°/s is a good average. A human driver reacts with up to 18°/s in dangerous traffic situations. Experiments were carried out with both values. It is interesting, that the higher change rate of the steering angle gives better results.

6 Fuzzy Control

In the realization of the controller, consideration of conventional controller techniques was deliberately not taken into account (nor of the establishment of a corresponding system model). In fuzzy control, the rule strategies of a human agent are transferred in the form of a sentence composed of linguistic variables over to a fuzzy controller. These linguistic variables have the form of words that are defined within a fuzzy set. The main advantage lies in the ability to immediately extract a heuristic from experiences. It also obviates the need for a mathematic process model. Instead, within the cognitive framework of the human agent having experience with a certain system, a fuzzy model manifests itself almost immediately [Kickert, Mamdani 1978]. To create a fuzzy controller for complex industrial processes, a human being imparting his expert knowledge would be required. For the mundane case of guiding a vehicle, consultation with a human being familiar with steering vehicles would be adequate to derive the fuzzy rules involved in the lateral control of a vehicle.

Other examples of the use of fuzzy control for controlling vehicles employ different approaches. How to regain control of a sliding or skidding car is the focus of [Altrock, Krause and Zimmermann, 1992]. [Sugeno et. al., 1989] used a fuzzy controller to control the angle of the front wheels and the speed of a model car by linguistic instructions given by an operator over a wireless transmission to the model car.

For the development of the fuzzy controller, a fuzzy compiler [INFORM 1992] was used consisting of a development shell, a debugger and a precompiler for ANSI-C. The source code can be linked to the main program as an ordinary procedure. Different fuzzy controllers were implemented based on this model. In the first case, only the degree of deviation from the planned value was used. In the second case, additional input of angular velocity was provided for the fuzzy controller. As expected, the second controller is more effective. It was this controller that was then used in the experimentation and the results of its performance are given here.

The fuzzy control sequence--fuzzyfication, inference, defuzzyfication is performed as described in most textbooks on fuzzy theory [Kickert, Mamdani 1987; Klir, Folger 1988; Kaufmann, Gupta 1991; Zimmermann 1991].

6.1 Fuzzification

The angle position of the corridor skirting boards in the video image is passed from the image processing to the fuzzy controller. From this, the angular deviation θ of the vehicle in the corridor can be determined. From the angle difference between the two pictures, the angular deviation velocity can be calculated.

 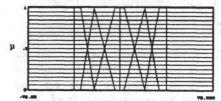

Fig. 4 The fuzzy sets for the variable lateral angular deviation θ and the attributes large, medium, small symmetrically grouped around the zero attribute.

Fig. 5 The fuzzy sets for the variable lateral angular deviation velocity $\dot\theta$ and the attributes large and small grouped symmetrically around the zero attribute.

Two fuzzy controllers using different input were developed and tested for lateral steering of a vehicle. For the first type of controller, only angular deviation θ from the nominal path was used as input; for the second type, the angular deviation velocity served as an additional input. As expected, the second type of controller produced better results. Only this controller will be presented.

The simple triangular fuzzy sets for the variables θ, and τ as optimized in the simulation and used in the experiments under description are shown in Figs. 4 to 6. The linguistic variables are grouped symmetrically around the centre.

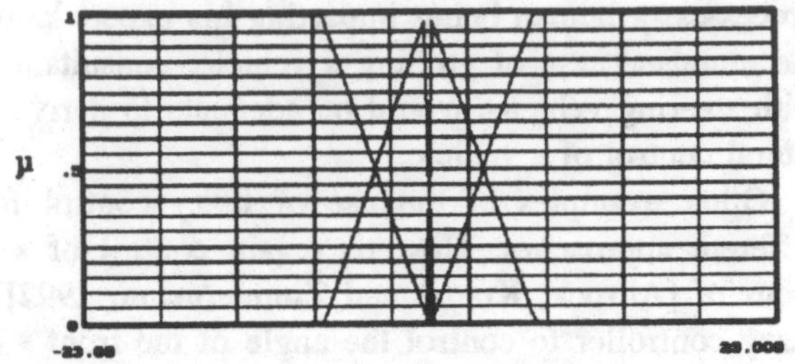

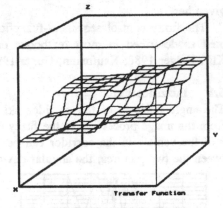

Fig. 6 The fuzzy sets for the steering angle τ and the attributes large and small grouped symmetrically around the zero attribute for left or right deviation.

6.2 Inference

Fuzzy inference requires the definition of an operator for aggregation and composition. There is a variety of operators and the type used depends on the context. In the preceding application, the MIN-operator was used for the aggregation and the PROD-operator was used for the composition without any compensation.

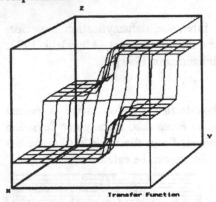

Fig. 7 3-dimensional transfer function without plausibility modification (degree of support). Inference rules are only weighted by zero and one. ($x = \theta$, $y = \dot{\theta}$, $z = \tau$)

Fig. 8 3-dimensional transfer function with modified plausibility in the intervall [0,1]. This transferfunction is used in the experiments. ($x = \theta$, $y = \dot{\theta}$, $z = \tau$)

The resulting transfer function is shown in Fig. 7. The origin of coordinates of the grahics lies in the centre of the cube. The angular deviation is drawn on the x-axis. The angular deviation velocity $\dot{\theta}$ is drawn on the y-axis. The steering angle is drawn on the z-axis. This transfer function shows steep gradients by even minor deviation from the planned course. A somewhat milder transition is more desirable. In order to attain this, a compensation (Gamma operator, also called plausibility) was introduced. Compensation implies the execution of a compensatory AND, just as it figures in our language and thought patterns [Zimmermann, Zysino

1980]. This shape is reached by introducing a plausibility degree for particular rules. Such a transfer function is shown in Fig. 8.

6.3 Defuzzification

The fuzzy inference delivers the output value as a fuzzy variable. For defuzzification the method "centre of area", often called "best compromise" is used. This well-known operation does not require any further explanation.

7 Results

Experiments with ATHENE were carried out in which the fuzzy controller developed in the simulation environment was tested. In the beginning, a velocity of 0.5 m/s was chosen for the vehicle, the steering angle change was limited to 0.32°/cycle. The velocity was then increased to a maximum of 1,5 m/s and the upper steering angle to 0.64°/cycle. During the experiments, the angles delivered by the fuzzy controller (x) and the actual angles of the vehicle (O) were logged and are shown in Figs. 9 to 13. The cycles of the control data transfer (110-170 ms) are numbered consecutively.

The angle and the offset of the vehicle at the starting point is the result of manual positioning. This coincidence

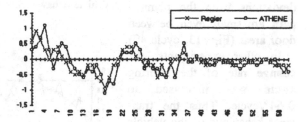

Fig. 9 The starting sequence of an experiment series, with the velocity 1.5 m/s and the change rate of steering angle set to 0.32°/cycle.

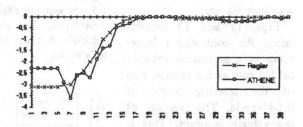

Fig. 10 Experiment with a velocity of 0.5 m/s and the change rate of steering angle set to 0.64°/cycle (upper limit). The vehicle meets the planned path with a minimal oszillation and follows it correctly.

makes it possible to test the transient behaviour of the vehicle. This phase is shown in Figs 9, 10, 11.

Doors in the corridor do not have skirting boards (edges). The "edge finder" recognizes the lower edge of the door, shifted in the image approx. 1.5° to the right of the skirting board. As a result, the planned path undergoes a sharp shift to the right. The fuzzy controller will therefore demand a positive steering angle to correct the shift. When recognizing skirting boards again, a backward shift is required. A disturbance of this kind is very useful to test the controller's stability (Figs. 11 and 13).

If the change rate of the steering angle is limited to 0.32°/cycle, more time is needed to reach the planned path. The limitation of the change rate hinders the controller in its work. Disturbances due to the lack of the skirting board in the areas of doors result in a minimal oscillation of the vehicle. The deviations from the planned path are small even between door areas (Fig. 13, cycle 40).

In other experiments, the change rate of the steering angle was increased to 0.64°/cycle. Thus the transient behaviour of the vehicle could be improved (Fig. 10). Initially, the velocity of the vehicle was limited to 0.5 m/s. The dynamic behaviour was very stable.

Figs. 12 and 13 demonstrate the controller's behaviour at a maximum velocity of 1.5 m/s and a change rate of the steering angle of 0.64°/cycle. The reaction of the vehicle is exact. This is shown in Fig. 11 where the vehicle's trajectory is stable but the angle of the steering wheel oscillates.

The peak at cycle 40 of Fig. 13 again represents a disturbance caused by a door.

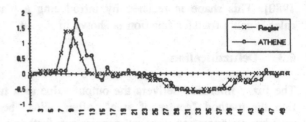

Fig. 11 Experiment with a velocity of 0.5 m/s and the change rate of steering angle set to 0.64°/cycle (upper limit). Even with a disturbance (at cycle 10) the vehicle behaves very stable.

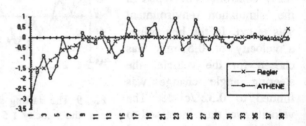

Fig. 12 Experiment with the maximum speed of 1.5 m/s and the change rate of steering angle set to 0.64°/cycle. The steering angle oszillates after the vehicle's start independently from the given controller's angle.

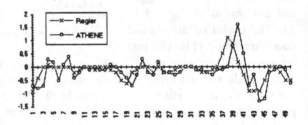

Fig. 13 Experiment with 1.5 m/s and the maximum change rate of the steering angle of 0.64°/cycle. The vehicles drive is not as stable as with the velocity of 0.5 m/s.

8 Conclusion

Users familiar with the basics of fuzzy logic can develop a fuzzy controller that will provide good results within a short time. No special mathematical skills are needed to understand the fuzzy operations.

9 Acknowledgements

The work described here was performed during my stay at the Universität der Bundeswehr München, Faculty of Luft- und Raumfahrttechnik, Institut für Meßtechnik headed by Prof. Dr. Graefe. I wish to thank K.-J. Wershofen who supported the work by providing the communication protocol for ATHENE, L. Tsinas for help with the image processing and J. Langbauer for his support in simulations and experiments. Thanks are also due to BMFT (Bundesministerium für Forschung und Technologie) and the German automobile industry working in conjunction with EUREKA-Projekts PROMETHEUS (Program for European Traffic with Highest Efficiency and Unprecedented Safety) without whose help the project would not have been possible and to the employment office.

References

von Altrock, C,; Krause, B.; Zimmermann H.-J. (1992): Advanced fuzzy logic control of a model car in extreme situations. *Fuzzy Sets and Systems* 48, 41-52.

Blöchl, B. (1992). Coarsely Grained Multiprocessor Image Processing System Coupled with a Transputer Network. In R. Trappel (Ed.), *Cybernetics and System Research '92-Vol.2*, 1423-1429. London: World Scientific.

Brooks, R. A. (1986): A Robust Layered Control System For A Mobile Robot. *IEEE Journal of Robotics and Automation*, VOL. RA-2, NO.1 March 1986.

Dickmanns, E.D.; Graefe, V. (1988). Dynamic Monocular Machine Vision. *Machine Vision and Applications 1*, 223-261. Munich, Germany.

Graefe, V. (1984): Two Multi-Processor Systems for Low-Level Real-Time Vision. In J. M. Brady, L. A. Gerhardt and H. F. Davidson (eds.): *Robotics and Artificial Intelligence*. Berlin: Springer.

Graefe, V.; Blöchl, B. (1991). *Visual Recognition of Traffic Situations for an Intelligent Automatic Copilot*. München: PROMETHEUS Workshop Oct. 1991. Unpublished manuscript available from: Institut für Meßtechnik LRT 6, Universität der Bundeswehr, Werner-Heisenberg-Weg 39, D-8014 Neubiberg, Germany.

Graefe, V., Kuhnert, K.-D. (1991): Vision Based Autonomous Road Vehicles. In: *Vision-Based Vehicle Guidance*. I. Masakio (Ed.). Springer, Berlin.

INFORM (1992). *fuzzyTECH 2.0*. Handbuch und Referenz. Aachen, Germany

Kaufmann, A. and Gupta, M. M. (1991). *Introduction to Fuzzy Arithmetic*. Van Nostrand Reinhold, New York.

Kickert, W.J.M. and Mamdani, E. H. (1978): Analysis of a fuzzy logic controller. *Fuzzy sets and Systems* 1. pp 29-44.

Klir, G. J. and Folger T. A. (1988). *Fuzzy Sets, Uncertainty, and Information*. Prentice Hall (UK) limited, London.

Kuhnert, K.-D.(1988): *Zur Echtzeit-Bildfolgenanalyse mit Vorwissen*. Dissertation, Fakultät für Luft- und Raumfahrttechnik der Universität der Bundeswehr München.

Nishiki-cho, Chiyoda-ku (1991). *Fuzzy Engineering toward Human Friendly Systems*. IFES Vol.1+2, Ohmsha LTD. Tokyo, Japan.

Sugeno, M.; Murofushi, T.; Mori, T.; Tatematsu, T.; Tanaka, J. (1989). *Fuzzy Algorithmic Control of a Model Car by Oral Instructions*. *Fuzzy Sets and Systems* 32 (1989). pp. 207-219.

Tsinas, L.; Graefe, V. (1992). *Automatic Recognition of Lanes for Highway Driving*. *IFAC/IMACS/IEEE/IUTAM workshop on Motion Control for Intelligent Automation*. Perugia, Italy.

Wershofen, K. P. (1992). *A Real-Time Multiple Lane Tracker for an Autonomous Road Vehicle*. Corfu: Euriscon.

Zapp, A. (1988). *Automatische Straßenfahrzeugführung durch Rechnersehen*. Dissertation, Fakultät für Luft und Raumfahrttechnik der Universität der Bundeswehr München.

Zimmermann, H.-J. & Zysino, P. (1980). *Zugehörigkeitsfunktionen unscharfer Mengen*. Berlin: DFG-Forschungsbericht.

Zimmermann, H.-J. (1991, 2nd Ed.). *Fuzzy Set Theory - and its Applications*. Kluwer Academic Publishers Group. Dordrecht, The Netherlands.

Robot Motion Coordination by Fuzzy Control

V. Khachatouri Yeghiazarians, B. Favre-Bulle

Institute of Flexible Automation
Technical University VIENNA

Abstract. Conventional Robot Motion Coordination is usually based on the tasks of path planning, coordinate transformation and the solution of the inverse kinematic problem. This concept is widely used in industrial applications, where pre-determined paths have to be accurately followed (cartesian path planning) or a Point To Point movement is to be executed without considering any defined cartesian path (high speed movement). The approach presented in this paper prepares a way between the pure cartesian and pure joint-coordinate based motion coordination. By implementing a rule base system for the generation of trajectories, the solution of the inverse kinematic problem can be avoided. In exchange, a Fuzzy control loop is closed between the actual position of the Tool-Center-Point, which is determined by the forward kinematics, and the joint drives. Following this strategy, the joint- and link-configuration can be controlled by a set of linguistic rules, which enables a new approach for the solution of robot body-collision avoidance problems and coordinate transformation problems in connection with redundant manipulators for future investigations. In this paper, a new rule based motion coordination concept is presented, and the properties of the method are demonstated by means of a simulation.

1 Introduction

Applications of Fuzzy Logic in the field of Robotics are reported in connection with Mobile Robots[1], Fuzzy Robot Axis Control[2], Robot Sensor Integration[3], Force Control[4] and Error Diagnosis Systems[5], just to mention a particular selection. Especially when considering the robot's evolution process from a robust, but insensitive handling machine towards an intelligent, sensor equipped manipulation mechanism, a clear trend towards achieving human abilities can be noticed. Further on, the range of application for robotic manipulators tends to shift from 'factory slave-tasks' towards more sophisticated activities, partly also dealing with human interaction. Just to mention an example, the field of rehabilitation robotics is continuously growing in Japan these days. For this trend of 'humanizing' the robots behaviour, the theory of Fuzzy Sets and the concept of rule based control systems seem to provide appropriate tools for making robotic action and interaction in non-industrial environments more flexible. In this paper, a new approach (AT Pat.Pending) for robotic motion coordination is introduced. The outstanding feature of this concept is the avoidance of the so called 'inverse kinematic problem', which is the transformation of world coordinates into robot coordinates. In conventional controller units, this problem has to be repeatedly solved in realtime for any given posture (position and orientation) of the Tool-Center-Point along the programmed path of the robotic manipulator, in order to determine the actual motor settings. Since this transformation may be very extensive, it forms a calculation-time bottleneck in real time motion coordination. The alternative approach of substituting the inverse coordinate transformation with a Fuzzy control loop will be discussed in this paper.

2   Fuzzy Control Systems

In this paper, no review of fundamental ideas behind Fuzzy Set Theory shall be given, since there is excellent literature on this subject [6]. Fuzzy Logic is the generalization of binary logic. Utilizing the tools given by the Fuzzy Set Theory, control loops can be closed making use of rule based algorithms. There are three main processes to be implemented in a Fuzzy controller, in particular *Fuzzification* (transformation of input variables into the Fuzzy domain), *Inference* (Combining the Rule Base with the variables) and *Defuzzification* (Retransformation of Fuzzy variables into signal domain). Fig. 1 shows the schematic structure of a Fuzzy Controller. The high degree of freedom for the design parameters of a Fuzzy Controller (rule base, observed process variables, appropriate fuzzy operators, defuzzification method etc.) shows the high need of human intuition in the design process of the basic fuzzy system.

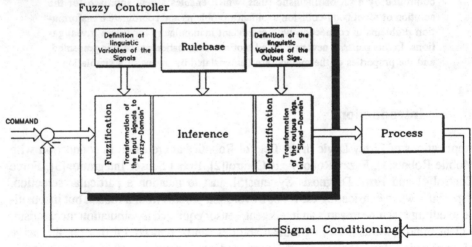

Fig.1. Fuzzy Controller

3   Robot Trajectory Generation

In conventional robot path planning units, usually two cases are distinguished in connection with trajectory generation:

o   Cartesian (World Coordinate) Trajectories
o   Joint (Robot Coordinate) Trajectories

In the first case, the calculation of a three dimensional path is done. Usually, the path is described by means of an analytical function in parametric representation

$$P_1(t)_{x,y,z,\phi,\theta,\psi} = F(t).$$

The parameter t then is set to zero at the beginning of the trajectory, and incremented each time interval to achieve the desired speed and acceleration profile. Thus a 'parameter-generator' produces values for t, which are substituted into the path equation

$$t_1 = PARGEN(t_{system-clock}).$$

The three-dimensional vector components, calculated from two subsequent path points

$$P_1(t) - P_{1-1}(t) = \Delta P_1(t)$$

show the spational direction vector.

If two parts of a trajectory have to be connected without intermediate stopping, the tangential speed components of the ending trajectory and the continuing trajectory must match. In this case, intermediate points, so called 'pass-points' have to be defined, which are 'passed by' in order to smoothen the path from the first to the second trajectory part.

The path equation generates n-DOF(Degree Of Freedom) tuples ($n_{max} = 6$) of cartesian coordinates

$$\vec{C}_1^{cartesian} = (x, y, \ldots, coord_n)_1$$

which are transformed into robot joint coordinates by means of coordinate transformation $\underline{A}$ and the solution of the inverse kinematic transformation $\underline{T}^{-1}$. $\underline{A}$ is calculated for offset shifts and rotations of the base- and tool systems of the robot

$$\vec{C}_1^{joint} = \underline{A} \cdot \vec{C}_1^{cartesian} \cdot \underline{T}^{-1}.$$

In case of Joint Trajectory generation, the path is planned in joint space

$$P_1(t)_{j_1,j_2,j_3,\ldots,j_n} = F(t).$$

The same parameter generator as in the cartesian case can be used to impose a speed and acceleration profile and to achieve the raw joint coordinate vector C_1. The remaining action to be done is to treat possible joint offsets by a transformation $\underline{B}$

$$\vec{C}_1^{joint} = \underline{B} \cdot \vec{C}_1.$$

Additional calculations are necessary, if trajectory overlapping is to be applied by means of 'pass-points'. This task is usually carried out by adding speed vectors of the 'fade-out' and 'fade-in' trajectory.

4   Trajectory Coordination by Fuzzy Control

To demonstrate an application example for rule based trajectory generation, the newly investigated principle of fuzzy controlled motion coordination shall be presented in the following. The task within this example is to move a robot's tool center point from point A to B by progressively adding displacement increments, determined by fuzzy control. In contrast to the conventional path planning principle, the trajectory generation is carried out in cartesian space without needing to calculate the inverse kinematics.

5   Bypassing the Inverse Kinematic Problem

Using the control scheme in Fig.2, the necessity of solving the inverse kinematic problem of the manipultor can be avoided by producing a feedback loop from joint position commands to the summing junction of a cartesian command position by means of a forward kinematic calculation.

Since the forward kinematic transformation is in general easier to solve than the inverse one, this part of the method usually needs a shorter calculation time for the kinematics. The Fuzzy controller itself is structured like in Fig.1. The rule base has to determine the reaction of joint increments caused by cartesian positioning error signals. Numerically summing up the joint increments starting from their initial position produces actual joint values, which are sent to the axis servo's.
Servo control is done conventionally by closing a loop around rotary angle encoders

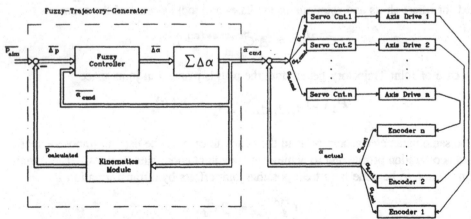

Fig.2. Trajectory coordination without calculation of inverse kinematic

and servo controller input. For future investigations, the servo controllers could also be implemented as fuzzy controllers [2].

6   A simple example

To gain experience for the layout of the rule base, a simple planar two arm configuration has been used (Fig.3).

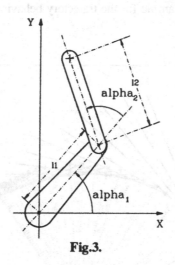

Fig.3.

The scheme for controlling this configuration is shown in Fig.4. The forward kinematic transformation is used to produce a feedback for the fuzzy controller. A crucial point in the controller design is the choice of the rule base. For the example configuration above, the working area of the manipulator is divided into several sub-areas, which are modelled as Fuzzy-Sets. For each fuzzy sub-area a set of fuzzy control-rules is defined. In order to avoid oscillations caused by switching between rule sets of neighboured areas, additional smoothing rules are used.

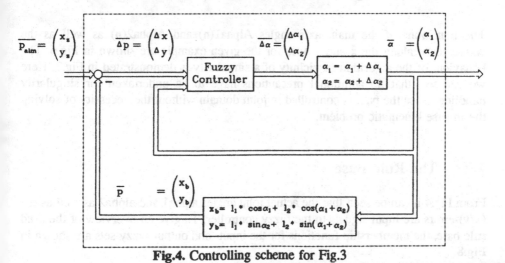

Fig.4. Controlling scheme for Fig.3

EXAMPLE

As for the simulation of this example, several choices for start and end points have been made, and the behaviour for the simple kinematics checked in depenency of certain rules. To give an example for the trajectory behaviour, Fig. 5 shows a movement sequence.

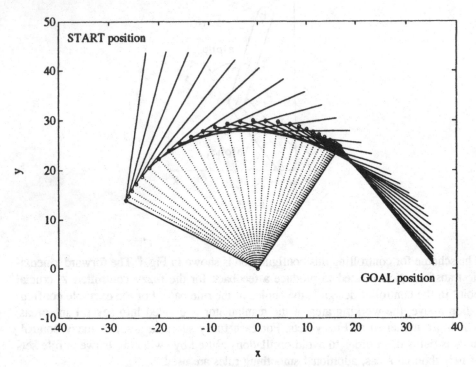

Fig.5. Fuzzy controlled movement sequence

The transitions of the main axis angles Alpha1(n) and Alpha2(n) as well as the convergence along the x and y axes for the given example are shown in Fig.6. The behaviour of the algorithm in vicinity of a singularity is demonstrated in Fig.7. Here we can see, that no additional precautions have to be undertaken for singularity handling, since the path is controlled in joint domain without the necessity of solving the inverse kinematic problem.

7 The Rule Base

From Fig.4 it can be seen that the actual joint angles alpha1 and alpha2 as well as the (x/y)-errors are input signals to the fuzzy controller. To give an overview of the used rule base, the membership functions for the input- and output fuzzy sets are shown in Fig.8.

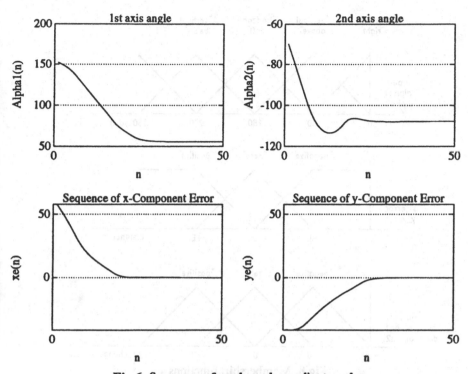

Fig.6. Sequence of angle and coordinate values

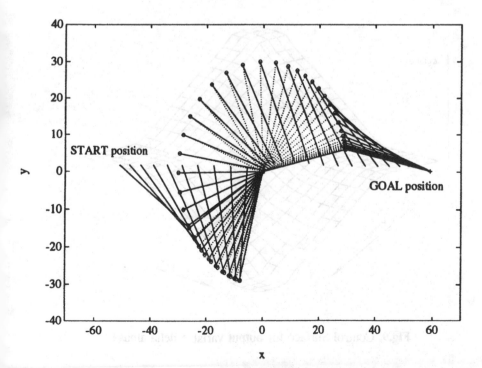

Fig.7. Movement through a kinematic singularity point

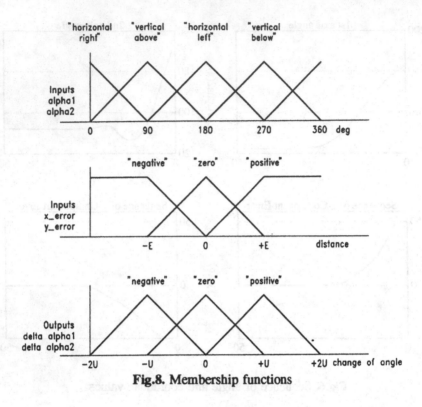

Fig.8. Membership functions

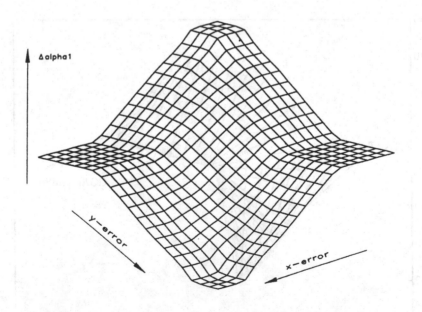

Fig.9. Control Surface for output variable delta alpha1

The scale factors E and U are dependent on link lengths and speed range. The rule base itself contains proportional rules associating cartesian position errors and joint angles in fuzzy domain with the output fuzzy sets of angluar joint-speed. Example: IF (alpha1 "Verical Above") AND [alpha2 ("Vertical Below" OR "Horizontal Right")] AND (delta x "Negative") THEN (delta alpha1 "Positive"). The AND-, OR- and NOT-operators are realized by pairwise minimum, maximum and order reversal respectively. Defuzzification is implemented by the centroid defuzzification method. The control surfaces in Fig.9 and Fig.10 show the output increments delta alpha1 and delta alpha2 of the discussed fuzzy controller that correspond to combinations of values of the two inputs x-error and y-error. The input variables alpha1 and alpha2 have the constant values 45deg and 315deg respectively. the values of the input variables x-error and y-error are varied from -1.4*E to +1.4*E in steps of 0.14*E.

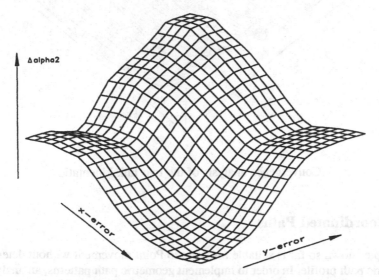

Fig.10. Control Surface for output variable delta alpha2

8   Convergence Considerations

The number of necessary steps for the travel from START to GOAL position is dependent on the START- and GOAL positions themselves. The following diagram has been gained from a convergence analysis. For this purpose, the START position has been varied by means of the joint angles Alpha1 and Alpha2. The number of steps necessary to converge to the GOAL position is drawn into z-axis direction.

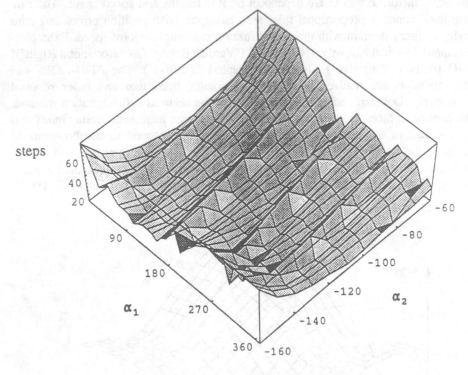

steps

60
40
20

-60
-80
-100
-120
-140
-160

α_1

90
180
270
360

α_2

Convergence analysis of the two link kinematics

9   Coordinated Paths

The principle shown so far is suitable for Point to Point movement without determined geometric path profile. In order to implement geometric path patterns, an analytical path formulation can be used, which describes the spatial trajectory

$$P_i(t) = F(t) .$$

An additional rule based fuzzy parameter generator produces a sequence of parameters

$$t_{i, fuzzy} = FPARGEN(t_{system-clock})$$

to impose a speed profile. The output of this generator determines the acceleration and deceleration process. Substituted into the geometric path equation, a set of START and GOAL points are calculated. Now, the Point to Point algorithm can be used to interpolate the fuzzy generated intermediate START and GOAL points of the geometric profile. In the case of redundant manipulator structures, the transformation of the cartesian GOAL position into the corresponding sets of joint coordinates is not unique. Also in this case, the proposed method shows a particular advantage.

10   Summary and Outlook

In the present paper, a new method for motion coordination has been introduced and discussed. Using a Fuzzy control loop, the calculation of the inverse kinematics of a robotic manipulator can be bypassed. By applying this solution, a number of advantages can be expected. First, the postures of the robot links in working space are controlled by means of linguistic variables, which gives advantage when treating obstacle avoidance problems. Second, the Fuzzy rule set can be extended for redundant manipulators, and give a handy solution for manipulator kinematics with high degree of redundancy. Since powerful microelectronic products for hardware support of Fuzzy applications are expected on the market, calculation speed limits will not be crucial for the realization of the method presented in this paper.

11   References

[1] R. Garcia,M.C. Garcia-Alegre, FUZZY LOGIC STRATEGIES TO CONTROL AN AUTONOMOUS MOBILE ROBOT, Cybernetics and Systems v 21 n 2-3 Mar-Jun 1990. p 267-276

[2] R. Tanscheit, E.M. Scharf, EXPERIMENTS WITH THE USE OF A RULE--BASED SELF-ORGANISING CONTROLLER, Fuzzy Sets and Systems v 26 n 2 May 1990 p 195-214

[3] R. C. Luo, MULTISENSOR INTEGRATION AND FUSION IN INTELLIGENT SYSTEMS, IEEE Transactions on Systems, Man and Cybernetics v 19 n 5 Sep-Oct 1990 p 901-931

[4] H. Kang, G. Vachtsevanos, AN INTELLIGENT STRATEGY TO ROBOT COORDINATION AND CONTROL, Proceedings of the IEEE Conference on Decision and Control v 4. Publ by IEEE 1990, IEEE Service Center, Piscataway, NJ, USA (IEEE cat n 90CH2917-3). p 2209-2213

[5] Y.Rong,M.Fang, H.S.Tzou,W.A.Gruver, MONITORING AND DIAGNOSIS OF ROBOT OPERATION THROUGH VIBRATION SIGNAL, Sensors, Controls, and Quality Issues in Manufacturing American

[6] H.J. Zimmermann, Fuzzy Set Theory and its Applications, Kluwer Academic Publishers, 1991

Fuzzy Control Schemes for Active Magnetic Bearings

Harri Koskinen

Machine Automation Laboratory, Technical Research Centre of Finland
Tampere, Finland

Abstract. Active magnetic bearings are used in applications where ordinary bearings meet difficulties. Due to nonlinear structure of active magnetic bearing system fuzzy logic control gives an attractive choice to control these systems. In this paper a fuzzy derivate gain adjusting method for active magnetic bearing control is presented. The method is based on the error and error derivate signals that are used as the basis of the fuzzy inference. With this method the effects of the vibration noise can be decreased.

1 Introduction

Active magnetic bearings (AMB) have proofed to be an attractive technology in applications where high rotating speeds or minimum friction and wear are demanded. The possibility to control the dynamic stiffness and damping is an other significant property in certain applications of active magnetic bearings. AMBs are nowadays serving in applications like turbo molecular pumps, flywheels for energy storage, reaction wheels for attitude control of artificial satellites, the main shafts of electric motors of tool machines and turbo compressors [1,2,3]. In spite of the tremendous progress made in recent years there are some difficult problems that still exist especially in the field of control of the AMBs.

The idea of applying fuzzy logic in dynamic system control has been studied since mid seventies [4] and during the past few years it has emerged to be one of the main application fields of the fuzzy set theory [5]. Although fuzzy logic control can be applied to well-defined systems, the best results have been reported from systems which are non-linear or ill-defined or when the parameter variations are excessive [5,6,7]. Fuzzy logic control has not been commonly employed in active magnetic bearing control yet, although some work has been done [8].

2 Active magnetic bearing system

The operation principle of the active magnetic bearing is seen in the Fig. 1. The rotor hovers between the magnets due to the attractive force of the electric magnets. The displacement sensor measures the rotor deviation from the reference position and this measurement is used in the controller. The control output determinates the currents that the power amplifier supplies to the magnets.

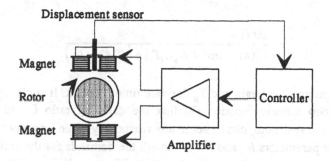

Fig. 1. The operation principle of the AMB

To understand the behaviour of the AMB system the linear force equation has to be studied. The linear force equation can be written

$$m\ddot{\delta} - K_b\delta = K_i i + F_e \qquad (1)$$

where m is the mass of the shaft, δ is the air gap between the shaft and the electric magnet, F_m is the magnetic force, F_e is the external disturbance force, K_i is the current gain and K_b is the position stiffness. It is important to notice that the position stiffness has always negative value, i.e. the magnetic bearing has a negative spring constant. Therefore the magnetic bearing has to be feedback controlled.

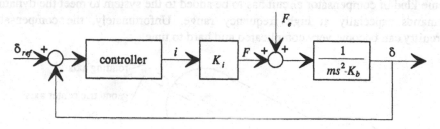

Fig. 2. The block diagram representation of the AMB system

Fig. 2 shows that the AMB system has two inputs, namely the reference air gap δ_{ref} (servo problem) and the external disturbance force F_e (regulator problem). In the practical systems the latter is far more important because in the normal operation the δ_{ref} is constant except during the start-up sequence. Usually the coordinate system for the air gap is determined so that $\delta_{ref}=0$ at the operating point. Because the force equation (1) has two states, a PD-controller should be sufficient. The control system usually contains a current controller too, but it has so small time constant compared with the rest of the system that it can be neglected. The transfer function for PD-controlled system can be written as

$$\frac{\Delta(s)}{F_e(s)} = \frac{1}{ms^2 + K_p K_i T_d s + \left(K_p K_i - K_b\right)} \tag{2}$$

where K_p is proportional gain and T_d derivate time constant. It is clearly seen from the closed loop transfer function (2) that the damping ratio ζ and the natural frequency of the system ω_n can be set to any reasonably value by proper selection of the controller parameters K_p and T_d. Although the demands for dynamic properties of the AMB system can be fulfilled with a PD-controller, a PID-controller is preferred because any steady-state error is not usually allowed [3].

However, some problems come up when ordinary PID-controller is used. The rotating shaft – even if it is carefully manufactured – has different geometric centre axis and actual rotating axis (Fig. 3). This means that the centripetal force fluctuates sinusoidal according to the rotating speed of the shaft and generates noise in the air gap measurement. Even a small eccentricity e in the shaft can cause troubles in the stability of the system due to the derivate term of the controller, because the PID-controlled system has high stiffness when the rotating frequency ω exceeds the system natural frequency ω_n (Fig. 4). Actually, the most hazardous situation occurs when the rotating frequency ω and the system natural frequency ω_n has the same value. Fortunately the natural frequency ω_n is usually much less than the rotating frequency of the rotor at the operating point and thus it can be neglected. Usually some kind of compensator circuit has to be added to the system to meet the dynamic demands especially at high frequency range. Unfortunately, the compensator circuitry can become very complicated and hard to tune.

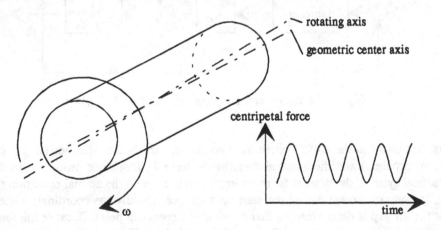

Fig. 3. Unbalanced shaft

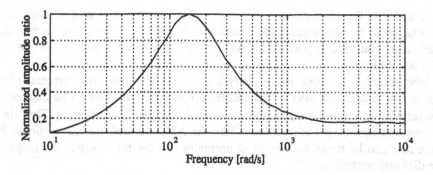

Fig. 4. The vibration amplitude dependence on the rotating frequency

3 Fuzzy Control Strategies of the AMB system

There are several potential possibilities to apply fuzzy logic in the control of an AMB. The first possibility is the use of a basic fuzzy logic controller (FLC) which can be considered as a traditional way of utilising fuzzy logic in a control system. The FLC in its traditional form was presented by Mamdani in his pioneer work [4]. This idea has widely adopted and it has been used to control almost everything from steam engine [4] and servo motor [9] to hydraulic position servo [7]. The structure of the FLC is presented in the Fig. 5. The fuzzy control rules in the FLC form the rule base and they have normally the form: "if x_1 is A_n and x_2 is B_n then u is C_n". The definitions for the fuzzy variables A_n, B_n and C_n are in the database.

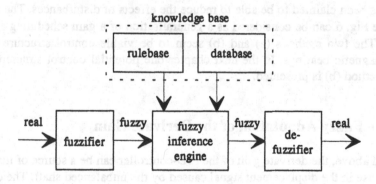

Fig. 5. The structure of the FLC

The fuzzy logic controller in its basic form has been the most successfully applied in systems which are non-linear, ill-defined or which are best controlled by human operator [5]. However, the systems, which can be mathematically modelled and controlled with the methods of the modern control theory, do not benefit much from fuzzy logic controller [7]. The tuning of the FLC is based almost merely on trial and error and it is therefore rather cumbersome. The stability and accuracy analysis of the FLC lacks comprehensive and practicable methods and tools. So, the traditional fuzzy logic controller is perhaps not the most suitable control structure for the active magnetic bearings.

Other methods to make use of fuzzy logic in control systems is presented in the Fig. 6. These methods have a common feature – they both have a PID-controller as the control basis and the fuzzy logic is an auxiliary part of the system. This kind of approach is advantageous in systems like the AMB control; the conventional PID-controller can be tuned and tested as normally and the fuzzy logic part helps with the difficult situations.

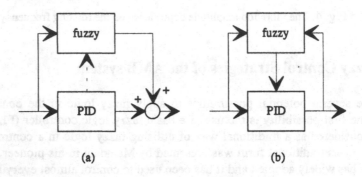

(a)   (b)

Fig. 6. Alternative fuzzy logic control structures

In the method (a) in the Fig. 6 fuzzy logic part of the system behaves as a compensator. This kind of structure has been used in a commercial PID-controller and it has been claimed to be able to reduce the effects of disturbances. The method (b) in the Fig. 6 can be considered as a generalisation of a gain scheduling adaptive control. The two methods (a) and (b) seem to be viable control structure for the active magnetic bearings. In the next chapter one potential control structure based on the method (b) is presented.

4   The Fuzzy Adaptation of the Derivate Gain

As stated above, the derivate gain of the PID-controller can be a source of instability due to noise in the displacement signal caused by the unbalanced shaft. The derivate control of a conventional PID-controller can be modified using fuzzy logic so, that the effects of the noise can be minimised [8].

The problem with the PID-controller is clearly seen from the frequency response (Fig. 7.): The gain is high at low and high frequencies. The high gain at low frequencies is desired because great stiffness improves the behaviour during the start-up and on the other hand the noise is no problem. But the increased gain at the high frequencies makes the system sensitive to the noise. So, the high gain at low frequencies should be maintained and, at the same time, the gain at the high frequencies should be decreased.

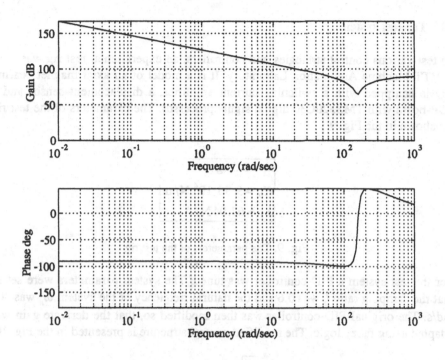

Fig. 7. The frequency response of the PID-controller

One method to decrease the gain at the high frequencies is to change the derivate gain as a function of the rotating speed. However, this method requires the rotating speed to be measured and it is not desirable to increase the amount of sensors. A more sophisticated approach is to use displacement and speed that are needed anyway. The decision how to alter the derivate gain using only displacement and speed is made with fuzzy rules [8].

The rules can be derived using the engineering knowledge about the system. The basic assumption is that when the absolute value of the speed is big the system is operating at high frequencies and the derivate gain should be reduced [8]. When the absolute value of the error (displacement) is big and the absolute value of the speed is also big the system is apparently heavily vibrating and the derivate gain should be reduced strongly [8]. The rule-base of the system is presented in the Fig. 8.

> Rule 1: IF *error* IS small AND *speed* IS small THEN *gain* IS big
> Rule 2: IF *error* IS small AND *speed* IS high THEN *gain* IS moderate
> Rule 3: IF *error* IS high AND *speed* IS small THEN *gain* IS big
> Rule 4: IF *error* IS high AND *speed* IS high THEN *gain* IS small

Fig. 8. The rule-base of the system

4.1 The test system

To test various control strategies and controllers, an experimental test rig was built in VTT Machine Automation Laboratory. It is a model of an axial magnet bearing that is equipped with a switching current amplifier, a displacement sensor and a DSP-board with TMS320C30 digital signal processor. The parameters of the test rig are shown in the Fig. 9.

$$m = 3.5\,\text{kg}$$
$$K_b = 9.93 \cdot 10^4\,\text{N/m}$$
$$K_i = 67.04\,\text{N/A}$$

Fig. 9. The parameters of the test rig

For the test system a PID-controller was tuned. The system parameters were set so that the damping ratio ζ was 0.6 and the natural frequency of the system ω_n was 300 rad/s. The original PID-controller was then modified so, that the derivate gain was adapted using fuzzy logic. The modified control structure is presented in the Fig. 10.

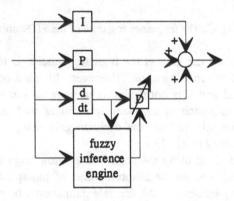

Fig. 10. The fuzzy derivate gain adjusting method

The controller was first simulated and then tested using the rules in the Fig. 8 and the membership functions in the Fig. 11 for *error* and *speed* (=*error derivate*). In the experiments the rudimentary min-max -inference method was used. The simulated amplitude response of the AMB system can be seen in the Fig. 12. The normal PID-controlled system has small amplitude both at low and high frequencies and an amplitude peak at the natural frequency. The fuzzy derivate gain adjusting controller has small amplitude at low frequencies but at high frequencies the amplitude is bigger than with normal PID-controller. As it is seen from the Fig. 12, the behaviour of the fuzzy derivate gain adjusting controller fulfils the expectations.

With the test system problems arise due to the short sampling time (0.167 ms) – great effort was needed to optimise the calculation of fuzzy part of the system in this tight time frame. The test rig used in the experiments gave results similar to those than simulations. However, the usefulness of this method can be found not until the tests with actual rotating machine are done.

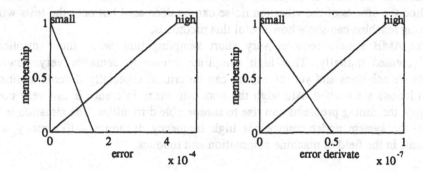

Fig. 11. The membership functions for the fuzzy variables

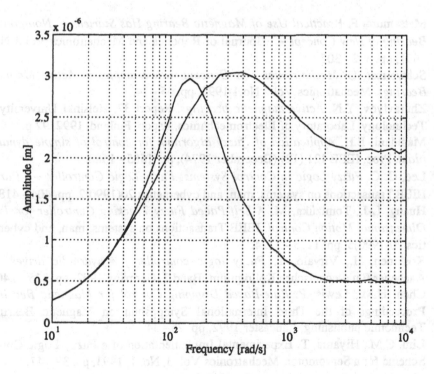

Fig. 12. The simulated amplitude responses

5 Concluding Remarks

In this paper a fuzzy derivate gain adjusting method for active magnetic bearing control is presented. The method is based on the error and error derivate signals that are used as the basis of the fuzzy inference. The simulations show that with this method the effects of the vibration noise can be decreased but only the tests with a rotating machine can show how useful this method is.

The AMB system requires very short sampling time when the controller is implemented digitally. This high sampling frequency demands very powerful hardware solutions and yet the timing can be critical especially if some additional calculations are needed. Although the max-min fuzzy inference is not very power hungry, the timing problems can rise to insuperable difficulties. The situation is bad when the system under control has high frequency dynamics – like many servo systems in the field of machine automation and robotics.

References

1. Matsamura, F. *Practical Use of Magnetic Bearing Has Started – Non-contact Bearing of New Concept – *. Journal of Robotics and Mechatronics Vol.3 No.4 1991, pp. 302 - 305.
2. Schweitzer, G. *Mechatronics – A Concept with examples in Active Magnetic Bearings*. Mechatronics Vol.2 No.1 1992, pp. 65 - 74.
3. Zhuravlyov, Y.N. *Active magnetic bearings*. Report 37, Helsinki University of Technology, Laboratory of Electromechanics, Espoo, Finland, 1992, 92 p.
4. Mamdani, E.H. *Application of fuzzy algorithm for control of simple dynamic plant*. Proc. IEEE 121 (12) (December 1974), pp. 1585 - 1588.
5. Lee, C. C. *Fuzzy Logic in Control Systems: Fuzzy Logic Controller – Part I*. IEEE Transactions on systems, man, and cybernetics 20(1990)2, pp. 404 - 418.
6. Huang, L-J., Tomizuka, M. *A Self-Paced Fuzzy Tracking Controller for Two-Dimensional Motion Control*. IEEE Transactions on systems, man, and cybernetics 20(1990)5, pp. 1115 - 1124.
7. Koskinen, H., Virvalo, T. *Fuzzy logic controller for hydraulic drives*. 10. Aachener Fluidtechnisches Kolloquium, Band 2, Germany 1992, pp. 225 - 240.
8. Chen, H.M., Lewis, P. *Rule-Based Damping Control for Magnetic Bearings*. Proceedings of the Third International Symposium on Magnetic Bearings, Technomic publishing, Lancaster 1992, pp. 25 - 34.
9. Lim, C.M., Hiyama, T. Experimental Implementation of a Fuzzy Logic Control Scheme for a Servomotor. Mechatronics Vol. 3, No. 1, 1993, pp. 39 - 47.

An Adaptive Fuzzy Control Module
for Automatic Dialysis

Silvio Giove (*), Maurizio Nordio (**), Alessandro Zorat (*)

(*)   Department of Computer and Management Sciences, University of Trento, Trento –
         ITALY
(**)   Department of Nephrology and Dialysis, Venice Hospital, Venice – ITALY

Keywords: fuzzy control, adaptive control, fuzzy rules, dialysis, nephrology

Abstract:

This paper presents a fuzzy adaptive control technique to automate the dialysis procedure.
The main points of this research are: use of a data base of fuzzy inference rules to formalize
the medical experience, use of pre-processed lookup tables to speed up the computation and
to allow for run-time changes of the inference rules, implementation of adaptive concept by
means a set of performance tables, and – finally – the smooth blending of different control
actions to obtain a single output action.

1.   Introduction

In biological terms, a dialysis procedure (i.e. the extra-corporeal filtering of the
blood stream) is a major perturbation of the patient's biological system that can
produce several undesired side-effects, the most important of which is the sudden
decrease of blood pressure that can cause dangerous hypotensive collapses.
Therefore, while a dialysis is being administered, several factors have to be carefully
monitored to achieve the desired effect, while containing as much as possible the
inception of undesired – and possibly dangerous – side-effects.
Typically, the procedure is carried out under direct medical supervision, with a
doctor manually adjusting a number of control variables in response to the dynamic
behaviour of several input variables continuously reported on a computer screen.
The doctor in charge of the dialysis apparatus usually tries to adjust the available

control variables as to satisfy some objectives, such as the rate of decrease of body weight, the rate of salt and other substances removal, etc.

Medical experience is of primary importance both to determine the objective functions that should be followed – functions that depend on several patient-specific factors – and to set the control variables correctly so that the desired functions as closely followed, all the while avoiding to reach undesirable or dangerous situations such as the sudden blood pressure drop mentioned earlier.

While medical experience is of paramount importance, the direct manual control of the dialysis procedure has the usual drawbacks implicit in every human activity: high costs, subjectivity of the decisions, possibility of errors due to the repetitiveness of actions, to lack of attention, to tiredness and so on. An automatic control mechanism that can autonomously adjust the control variables based on the observed input variables would be very useful in overcoming these problems.

With the availability of ever more powerful hardware, it is now possible to develop an automatic feedback control for dialysis at a viable cost. Recent studies have been presented investigating the feasibility of such a controller. For example, in [1] and [2] a closed-loop control is proposed based on a statistical description of the interaction between the biophysical dynamic model of a nephropatic and the dialysis machine.

However, the medical knowledge on this subject is mostly based on the experience, rather than on a precise mathematical model, and thus it is expressed primarily on qualitative terms, such as "when the salt concentration is low and dropping rapidly, drastically reduce the instantaneous rate of fluids extraction". This suggests that the automatic controller be based on fuzzy rules such as the one above. In addition, the controller ought to be adaptive, so that the meaning of "low concentration", "drastically reduce" etc. can be adjusted dynamically to better follow the desired global behaviour.

In this article an adaptive fuzzy logic controller for dialysis procedures is presented. It is based on on-line fuzzy rules and it uses an automatic self-regulating action, as described in [10], to dynamically correct its data rule set.

Automatic control of the dialysis procedure is particularly suited for this kind of fuzzy adaptive control, since:

1) The dynamics of the biological system are poorly known; furthermore, as shown in a number of studies [3], [4], [5], the evaluation of the current "status" would require the measure of non-accessible variables, such as the intra-cellular concentration of salt.

2) As mentioned earlier, the control is primarily based on medical experience, which is more naturally expressed by rules defined in linguistic terms, rather than by means of the mathematical formulas that are usually used in classical PID controllers. In addition, the rules embodying the medical knowledge, are often not very precise (different doctors can express different judgements requiring a preliminary mediation).

3) The (unknown or poorly known) parameters and dynamics of the biophysical

system can vary during the dialysis, thus requiring that the rules be modifiable at run-time.

Fuzzy logic can take into account the subjectivity of the available knowledge and the poorly-defined or not-so-well-assessed inferences, while the adaptive action can use the necessary (implicit or not) modification of rules (for example, adding or eliminating rules, as in [9]) to dynamically adapt to the changes that take place during the dialysis.

2. Description of the Model

The hemodialysis procedure aims to filter the blood of a nephropatic patient by extracting blood from the patient veins by means of a set of needles, routing it through a particular filter acting as a blood purifier, and re-inserting it into the patient blood stream. During the administration of the hemodialysis it is necessary to reduce the body weight through a liquid extraction procedure, starting from its initial value and ending with a final value, characteristic of each patient. The negative blood pressure gradient thus created draws a liquid solution from intra-cellular and interstitial remote compartments and the resulting fluid unbalance can produce an undesirable effect known by the term "hypovolemia" which is a drastic fall of blood volume leading to pressure impairments.

Blood pressure variations depend directly on blood (or plasma) percentile volume changes (volemia) and hence blood pressure can be controlled indirectly by acting on the blood volume changes, i.e. by varying the instantaneous fluid extraction rate. However, other physiological conditions can induce pressure variations as well, and therefore blood pressure control cannot be achieved solely by blood volume monitoring.

In general, at least two variables have to be measured and controlled to avoid undesired effects on the patient: *volemia* and *blood pressure*. Presently, only systolic pressure is considered. The control can be performed mainly by varying some control variables, such as the liquid flow rate (difference between the out-flow – the so called "ultrafiltration" – and the in-flow) and the dialysate sodium concentration. In this paper, only the liquid flow action is considered, since the sodium control is a more recent procedure (in human control) and it is neither well known, nor much used in practice. No logical difficulty arises to extend the control procedure also to sodium concentration.

Several other state and control variables could be taken into account and their addition poses no problem at the logical level; at the implementation level the additional variables would require longer computation time but the sampling rate of the system is measured in tens of seconds and hence there is ample time available even for a small computer. More important is the question of whether additional variables are necessary for the correct behaviour. This question will be addressed during the forthcoming experimental phase of the fuzzy adaptive controller here

described.
At present the variables used are:

INPUT VARIABLES:
Blood pressure error
Blood pressure error rate
Volemia error
Volemia error rate

CONTROL VARIABLE:
Liquid flow variation, that is the instantaneous fluid rate, the difference between out-flow (the so called "ultrafiltration flow"), and the in-flow (that is the infusion flow, supplied by an external machine). Initially, at time 0, the in-flow is set to 0, while the out-flow is set to the value:

$$Uf(0) = \frac{W_i - W_f}{T} \quad \text{where:}$$

W_i is the initial body weight,
W_f is the desired final body weight,
T is the total time set for the dialysis procedure (typically 3–4 hours).

Obviously, both W_i and W_f are known and depend on the patient.
During the dialysis procedure, the control adjusts Lr if pressure and volemia errors are detected. The variation of the control variable needed to compensate for the errors are the bases of "medical experience", an art which is not easily defined by a mathematical formula, but that can be conveniently formalized by inference rules typical of fuzzy logic.

The rules can be compactly expressed as a multi-dimensional matrix for each output variable, with one dimension for each input variable; each dimension is indexed by the linguistic values that the corresponding variable can assume (e.g. NL, NS, ZE, PS, and PL). The values of the matrix represent the value that is to taken by the corresponding output. While in theory this would result in having to fill every position of the matrix, in practice not all combinations are to be examined: it is sufficient to cover the universe set, i.e. for every numerical (that is, "crisp") combination of possible input values in the considered range, at least one rule must be have its antecedent matching with percentage greater than zero (if not so, some alternative technique can be used; see [11]).

For this specific application, time is not a limiting factor, since sampling frequency is low (about thirty seconds), and hence computational problems do not arise. However, computation time could be traded off for storage, by pre-computing a lookup-table storing the values of the "crisp" actions above that should be taken for each input value combination, after an appropriate quantization of the input variables (see - for example [6]).

The lookup-table approach is also convenient for inclusion of a simple mechanism that adjusts the inference rules at run-time to adapt to drifts between the expected and observed values of the input variables, as detailed in the following paragraph.

In the case here considered each rule has four antecedents; the number of cases to be considered is relatively small and hence the problem is computationally tractable. However, should it be necessary to consider more input variables, the "size" of the problem grows very rapidly, in which case it is desirable to divide the problem into separate subproblems and then "merge" their solutions. While not strictly necessary for this particular case, the technique for problem subdivision and merging is here presented as a concrete example of the general case.

The control problem is thus divided into two separate subproblems: the first is related to the blood pressure control, while the second is related to volemia control. As described later on, a lookup table is used to generate an action (liquid flow variation) in response to the pressure inputs, while a second lookup table computes an action (again for the liquid flow variation) in response to the volemia inputs. The two actions are then combined to obtain a single output action, as explained in greater detail in [15]. Basically, a first-level algorithm takes the two values and combines them into a final value in such a way as to satisfy some safety conditions, such as the necessity that the output value be within some pre-established bounds and discounting the action for the volemia control when safety is compromised (when blood pressure control becomes the primary interest).

Let then the input and output variables be quantified into appropriate ranges of values and let P, Pr, V, and Vr be the *discrete* values taken by the corresponding input variables. Finally, let T be the time taken by the dialysis procedure be discretized into N sampling intervals, $\{t, t+1\}$, for $t = 0,...,N-1$, each of unit length.
In the following, let $L_p[\ ]$ and $L_v[\ ]$ denote the look-up table for the pressure and volemia, respectively.

For a given value of the blood pressure error P and the pressure error rate Pr, the action to be performed for the pressure is given by $L_p[P, Pr]$. Similarly, the volemia action is given by $L_v[V, Vr]$.

The two actions (on the single control variable Lr) are then combined by a suitable function merge ( ) to compute the final action that is to be actually performed, that is: $Lr = $ merge $(L_p[P, Pr], L_v[V, Vr])$ in such a way that the resulting Lr is such that:

$$\min (L_p[P, Pr], L_v[V, Vr]) <= Lr <= \max (L_p[P, Pr], L_v[V, Vr]).$$

When both P and Pr are relatively large, safety could become a concern, since rapid pressure drops could be dangerous to the patient. In this case, medical practice discounts the volemia readings and concentrates on the task of returning the blood pressure to its reference value, that is, actions are undertaken to reduce P and Pr.

Analytically, this can be done by defining merge( ) as follows:

$Lr = \text{merge } (L_p[P, Pr], L_v[V, Vr]) = \beta_1 \, L_p[P, Pr] + \beta_2 \, L_v[V, Vr],$

where $\beta_1 = 1 - (1 - \mu)\exp(-P + Pr)$ and $\beta_2 = 1 - \beta_1$

and μ is a fixed threshold value assigned by the doctor in charge and reflecting the relative importance of the blood pressure to the volemia.

So far, the determination of the action to be applied is very straightforward, the only difficulty being the selection of "good" $L_p[P, Pr]$, $L_v[V, Vr]$, and merge( ) for the specific application at hand. In the next section an adaptive mechanism is introduced to allow run-time modification of the lookup tables, in response to the "performance" of the previous actions.

3.   Adaptive Action

Since the lookup tables will now be allowed to change dynamically, a superscript (t) will be used to denote the contents of the lookup tables at time t, as follows: and

$L_v^{(t)}[\ ]$. Similarly, an input x read at time t will be denoted by $x^{(t)}$.

As in the previous section, the control variable Lr is then computed by merging the two values obtained by the value of the tables *at time t*, that is:

$$Lr = \text{merge } \left(L_p^{(t)}\left[P^{(t)}, Pr^{(t)}\right] L_v^{(t)}\left[V^{(t)}, Vr^{(t)}\right] \right).$$

Notice that now both the lookup tables and the input values are those at time t, as indicated by the (t) index.

The action that is performed at time t is then that indicated by Lr, which will cause a (possibly null) variation of the liquid flow so that the value of the pressure and volemia parameters at the next sampling point ought to be exactly those expected. At time $t+1$ the inputs are sampled again and the corresponding errors between measured values and reference values are computed, yielding $P^{(t+1)}$, $Pr^{(t+1)}$ and $V^{(t+1)}$, $Vr^{(t+1)}$. If these were zero – or very small – the action that was performed at time t turned out to be a "good" action and hence the lookup tables that caused that choice should not be modified. Hence: $L_p^{(t+1)}[\ ] = L_p^{(t)}[\ ]$ and $L_v^{(t+1)}[\ ] = L_v^{(t)}[\ ]$

If, however, the differences between observed and reference values turn out to be significant, the action taken was ill-advised and thus the entry $P^{(t)}$, $Pr^{(t)}$ of lookup table $L_p^{(t)}[\ ]$ and the entry $V^{(t)}, Vr^{(t)}$ of lookup table $L_v^{(t)}[\ ]$ must be updated so that the next time around the expected error will be reduced, or eliminated.

The adequacy of the rule set for those specific input values can be measured by a "performance table" as suggested in [10], relating the components of the overall error (that is, error value and rate) to a corrective value to be added to the respective element of the look-up table. In the case examined here, the performance table consists of two two-dimensional tables $Perf_p[\ ]$ and $Perf_v[\ ]$; the look-up table can then be updated as follows:

$$L_p^{(t+1)}\left[P^{(t)}, Pr^{(t)}\right] = L_p^{(t)}\left[P^{(t)}, Pr^{(t)}\right] + Perf_p\left[P^{(t+1)}, Pr^{(t+1)}\right] \quad \text{and similarly for}$$

$$L_v^{(t+1)}[\ ].$$

The decision of whether to apply the correction or not can be crudely based on a fixed threshold of the value of the expected and observed value of $P^{(t+1)}, Pr^{(t+1)}$ and $V^{(t+1)}, Vr^{(t+1)}$. However, a better strategy is to apply the correction to $L_x^{(t)}[\ ]$ proportionally to the difference between $Lr^{(t)}$ and $L_x^{(t)}\left[X^{(t)}, Xr^{(t)}\right]$ for $x \in \{p, v\}$.

That is, a positive, symmetric and differentiable function

$$\gamma_x^{(t)}\left(Lr^{(t)} - L_x^{(t)}\left[X^{(t)}, Xr^{(t)}\right]\right), \text{ for } x \in \{p, v\}, \text{ is defined such that:}$$

$$\gamma_x^{(t)}(0) = 0 \quad ; \quad \lim_{z \to \infty} \gamma_x^{(t)}(z) = 0 \quad ; \quad \dot{\gamma}_x^{(t)}(z) > 0 \text{ for } z > 0$$

For example: $\gamma_x^{(t)} = \exp\left[-\dfrac{\{Lr^{(t)} - L_x(X^{(t)}, Xr^{(t)})\}^2}{\sigma^2}\right]$, where $\sigma > 0$ is a constant.

The look-up table is thus continuously updated as follows:

$$L_x^{(t+1)}\left[X^{(t)}, Xr^{(t)}\right] = L_x^{(t)}\left[X^{(t)}, Xr^{(t)}\right] +$$

$$+Perf_x\left[X^{(t+1)}, Xr^{(t+1)}\right] \cdot \gamma_x^{(t)}\left(Lr^{(t)} - L_x^{(t)}\left[X^{(t)}, Xr^{(t)}\right]\right)$$

At the end of the first dialysis, for every patients the lookup-tables are stored into an appropriate file; the last updated version will be used in next dialysis.

4. Definition of Fuzzy Sets, Rules, and Tables

For the input variables the following fuzzy trapezoidal sets are used to quantify the natural language attributes:

Blood Pressure Error	(P)	[mmHg]:			
Negative (N)		-60.0,	-40.0,	-20.0,	0.0
Zero (Z)		-10.0,	0.0,	10.0,	30.0
Positive (P)		20.0,	30.0,	60.0,	110.0

Blood Pressure Error Rate	(Pr)	[mmHg/min]:			
Negative High (NH)		-1.0,	-0.8,	-0.6,	-0.4
Negative Low (NL)		-0.5,	-0.38,	-0.25,	-0.17
Zero (Z)		-0.25,	-0.13,	0.13,	0.25
Positive Low (PL)		0.17,	0.25,	0.38,	0.5
Positive High (PH)		0.4,	0.6,	0.8,	1.0

Blood Volume Error	(V)	expressed in percentage [%] with respect to its initial value:			
Negative High (NH)		-10.0,	-10.0,	-8.5,	-6.0
Negative Medium (NM)		-7.0,	-5.0,	-3.0,	-1.5
Negative Low (NL)		-2.5,	-1.0,	-0.5,	0.0
Positive (P)		0.0,	0.5,	1.0,	2.0

Blood Volume Error Rate	(Vr),	[%/min]:			
Negative High (NH)		-4.0,	-2.5,	-2.0,	-1.5
Negative Low (NL)		-2.0,	-1.0,	-0.5,	0.0
Zero (Z)		-0.3,	-0.1,	0.1,	0.3
Positive Low (PL)		0.0,	0.1,	0.3,	0.5
Positive High (PH)		0.2,	0.3,	0.7,	0.8

Liquid Flow rate	(Lr)	[l/h]:			
Negative High (NH)		-1.5,	-0.9,	-0.8,	-0.7
Negative Low (NL)		-0.8,	-0.5,	-0.4,	0.2
Zero (Z)		-0.3,	-0.05,	0.05,	0.1
Positive Low (PL)		0.0,	0.2,	0.4,	0.6
Positive High (PH)		0.4,	0.7,	0.8,	0.8

The following sets of rules are chosen:

P / P_r	N	Z	P
NH	PH	PL	NL
NL	PH	Z	NH
Z	PL	NL	NH
PL	Z	NH	NH
PH	NL	NH	NH

Pressure rule table

V / V_r	NH	NM	NL	P
NH	PL	PH	PL	Z
NL	PL	PL	Z	NL
Z	PL	Z	NL	NH
PL	Z	NL	NL	NH
PH	NL	NL	NH	NH

Volemia rule table

6.   Conclusions and Future Work

Some improvements are possible to the present version of the controller.
Additional variables can be taken into account, the main one being the sodium control – which will be certainly considered in the next implementation.
Moreover, since time is not a limiting factor for this particular application, the quantizing procedure implicit in the construction of the look-up tables could be substituted by the computation of the output actions directly from the "analogue" values of the variables, by means of center-of-gravity techniques, or - alternatively - by the simplified algorithm described in [9].

Other corrective actions could be used, as - for example - the ones described in [12], where the second rate of error is also considered to select the adaptive action. Also, different adaptive actions can be taken into account; in particular those based on modification of the procedure described here and in [9], [13].

Since volemia will change during the dialysis procedure, the problem here presented can be viewed more as a "servo-regulating" problem, rather than a classic "control-regulating" problem. Hence, the reference values that are to be achieved in future times ought to be considered, besides the values obtained in the past (see [16]).

Finally, the assignment of performance table values is a totally subjective tool, and become very difficult if the dimension of the matrix is large. To reduce this problem, it can be preferable to adopt a different approach, as the one described in [9], and the improved version described in [14] and adopting the simplified computational procedure proposed by Wu et al. [14] to speed up the "polling" of the Rules Data Base; our plans are to develop an alternative version of the control by introducing an automatic fuzzy adaptation mechanism to select the most significant rule and using a PDI discrete controller as input to the system.

Bibliography

[1]   Santoro A., Spongano M., Mancini E., Muratori A., Rossi M., Paolini F., Zucchelli P.: "Prediction of dialysis induced hypotension by means of continuous blood volume, blood pressure and heart rate monitoring", Nephrol. Dial. Transplant, 7 (7), 1992.

[2]   Santoro A., Paolini F., Rossi M., Mancini E., Spongano M., Bosetto A., Zucchelli P.: "Sistema di controllo automatico degli andamenti intradialitici del volume ematico", Atti del 32. Congresso Nazionale della Soc. Italiana di Nefrologia, Monduzzi Ed., Bologna, pp. 1139-1143, 1991.

[3]   Kimura G., Van Stone J.C., Bauer J.H., Keshaviah P.R.: "A simulation study on transcellular fluid shifts induced by hemodialysis", Kidney Int, 24, pp. 542-548, 1983.

[4]   Man N.K., Peticlerc T., Tien N.Q., Jehenne G., Funck-Brentano J.L.: "Clinical validation of predictive modelling equation for sodium", Artif. Org. 9, pp. 150-154, 1985.

[5]   Thews O., Deuber H.J., Hutten H., Schulz W., "Theoretical approach and clinical application of kinetic modelling in dialysis", Nephrol. Dial. Transplant, 6, pp. 180-192, 1991.

[6]   Ollero A., Garcia-Cerezo A.J.: "Direct digital control auto-tuning and supervision using fuzzy logic", Fuzzy Sets and Systems, 30, pp. 135-153, 1989.

[7]   Buckley J.J.: "Fuzzy I/O controller", Fuzzy Sets and Systems, 43, pp. 127-137, 1991.

[8]   Kickert W.J.M., Mamdani E.H.: "Analisys of a fuzzy logic controller", Fuzzy Sets and Systems, 1, pp. 29-44, 1978.

[9]   Wu Z.Q., Wang P.Z., Teh H.H.: "A rule self-regulating fuzzy controller", Fuzzy Sets and Systems, 47, pp. 13-21, 1992.

[10] Shi-Shang Jang, David Shan-Hill Wong, Chau-Kuang Liau: "On-line/off-line optimization of complex processes using a linguistic self-organized optimizing control scheme", Fuzzy Sets and Systems, **47**, pp. 23-33, 1992.

[11] Koczy L.T.: "Reasoning by analogy with sparse fuzzy rule bases", Proceedings of KFMS SPring Conference '92, **2**, (1), 1992.

[12] Maeda M., Murakami S., "A self-tuning fuzzy controller", Fuzzy Sets and Systems, **51**, pp. 29-40, 1992.

[13] Zhong He, Shaohua Tan, Chang-Chieh Hang, Zhuang Wang, "Control of dynamical processes using an on-line rule-adaptive fuzzy control system", Fuzzy Sets and Systems, **54**, pp. 11-22, 1993.

[14] Wu Z.Q., Wang P.Z., Teh H.H.:"A simplified fuzzy controller algorithm and its hardware implementation", 2nd International Joint Workshop on Fuzzy Logic and Neural Networks NASA/Johnson Space Center, Houston, TX (April 1990).

[15] Giove S., "Distribuited hierarchical adaptive fuzzy control", in progress.

[16] Giove S., "Feed-forward fuzzy adaptive controller", in progress.

A Combination Scheme of Artificial Intelligence and Fuzzy Pattern Recognition in Medical Diagnosis[1]

Ludmila I. Kuncheva[1], Roumen Z. Zlatev[2], Snezhana N. Neshkova[1], Hans Gamper[3]

[1] Central Laboratory of Bioinstrumentation and Automation,
Bulgarian Academy of Sciences, Acad. G, Bonchev Street, Bl. 105, 1113 Sofia, BULGARIA,
[2] United Scientific and Research Institute for Aviation Medicine, Sofia, BULGARIA
[3] Department of Medical Computer Sciences, University of Vienna,
Währinger Gürtel 18-20, A-1090 Vienna, AUSTRIA

Abstract. The statement advocated here is that a trainable scheme resembling some elements of physician's decision making process could lead to an increased classification accuracy. Problems are considered which are difficult to cope with from both pattern recognition and AI point of view. A combination scheme is proposed which consists of an AI preprocessing part and a pattern recognition part delivering the final decision. The idea is to detect and then to aggregate the strengths of positive and negative evidence for a given medical hypothesis through a trainable nonlinear function. A possibility for application to CADIAG-2 is considered. An example from aviation medicine is presented which demonstrates the classification accuracy of 76.6 % of the proposed scheme versus 64.1 % obtained with linear discriminant analysis.

1 Introduction

Although pattern recognition and artificial intelligence (AI) branches have usually been confronted and have developed in isolation from each other, the recent tendency is to try to combine their tools [1]. Their common difficulties and challenges arise from the common classification task [2]. In fact, neural networks which are attributed to both fields are an example of such common classification tools. The main emphasis, however should be put not on the common tools but on the integration of these different approaches. Problems which appear cumbersome for the physician are naturally difficult to interpret in terms of AI (the well known knowledge acquisition bottleneck). On the other hand, a sufficient sample which would allow to use classical Bayes' decision theory is never available. Therefore, the

---

[1]This work was partially supported by the Research Contract НИ-МУ-2-ИН/93 with the Young Scientists Foundation.

key to solving these problems is the combination of the vague expert information on the problem with that contained in the available data set.

Fuzzy set theory is usually applied to pattern recognition trying to include heuristic elements or expert opinions in the classification paradigm and thus to increase the classification accuracy or to handle special kinds of data which defy statistical interpretation. Some heuristic fuzzy pattern classifiers for medical diagnostic tasks are presented in [3,4,5].

This paper proposes a combination scheme which includes an AI preprocessing part and a fuzzy pattern recognition part delivering the classification decision. The underlying idea is to replace the classical scoring sum used to aggregate positive and negative evidence for a diagnosis with a more sophisticated technique. Due to this complication an increase in the classification accuracy is expected.

Section 2 describes the configuration of the combination scheme. Section 3 considers the AI preprocessing part in details, and Section 4 the pattern recognition one. Section 5 contains a comment on the possible application of the scheme to CADIAG-2. An example from aviation medicine is presented in Section 6.

2 Configuration of the Combination Scheme

The main idea is to detect and then to aggregate positive (PRO-) and negative (CON-) evidence for a given hypothesis, resembling in this way human decision making process in medical diagnostics. In his routine practice clinician is always faced with the situation when two or more tests or observations yield conflicting results. The way he copes with this case can hardly be expressed by simple weighted summation or by crisp rules. The widely used scoring sums imply just this inhibitory relationship of positive and negative strengths. On the other hand, heuristically designed rules with predefined certainty could also fail in a difficult contradictory case.

Instead, a combination of the two opposite strengths (expressed in the form of degrees of membership μ_{pro} and $\mu_{con} \in [0,1]$) is proposed through a trainable nonlinear function f (μ_{pro}, μ_{con}) which should vary monotonically between the four boundary values presented in table 1. Values 0 for μ_{pro} and μ_{con} denote no evidence while values 1 indicate evidence with maximal strength. Value 0 for f means "all evidence points at rejection of the hypothesis" and value 1 - for its acceptation.

Table 1. Boundary points for f (μ_{pro}, μ_{con}) corresponding to medical reasoning

μ_{pro}	μ_{con}	f	Reasons for rejection
0	0	0	Lack of positive evidence
0	1	0	Strong negative evidence
1	0	1	-
1	1	0	Ambiguous data

The main configuration of the scheme for one hypothesis is presented in fig 1

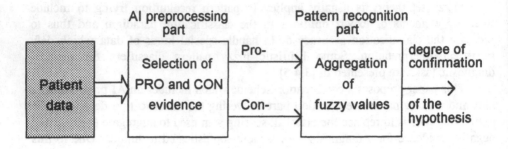

Figure 1. The combination scheme

Patient data is characterized by the feature set $\mathbf{X} = \{\ X_1, \ldots, X_n\ \}$ where X_i, i = 1,...,n, denote symptoms, signs, measurements, laboratory test results, clinical findings, etc. Each patient is presented as a vector of feature values $v^T = [x_1,...,x_n]$. Due to the embodying an AI preprocessing part, missing values in patient data are supposed to not affect the system performance. This fact is highlighted because pattern recognition techniques usually require full data sets which is hardly obtainable in medical diagnostic tasks. Let $D = \{\ D_1,...,D_M\ \}$ be the set of diseases for a medical diagnostic task under consideration. Although this configuration of the scheme computes the support for one hypothesis at a time it may be used in the respective context both at the initial stage of rising diagnostic hypotheses and in the ensuing differential diagnostic tasks. In the former case the hypothesis is D_i versus $D\backslash\{D_i\}$, while in the latter D_i versus D_j, $D_i, D_j \in D$.

3 AI Preprocessing Part

The particular implementation of the scheme depends on the form in which the AI preprocessing part yields the pro- and con- evidence. Among many possible forms the following three are considered.

(i) PRO- and CON- evidence is described directly in terms of initial features. If the value (or presence/absence) of a feature votes for the hypothesis this value is directed towards PRO- output of the AI preprocessing part, and if the value points at rejecting the hypothesis it is directed towards the CON- output, respectively. If the value does not definitely vote for or against the hypothesis it is not exported at all. As an example, let the current hypothesis be "hypertonia". If the systolic blood pressure is above 140 the exact value will be directed trough PRO- output while CON- output will yield "missing" for this feature. If it is below, e.g., 100, the CON- output will transmit it. For a value in the range (100, 140) both outputs will yield "missing".

Along with the initial features the output vectors contain several positions for evidence items obtained via combination of features (e.g., in syndromes) specific for that particular task.

This form of presentation of PRO- and CON- evidence exempts the AI preprocessing part from the need to attach numerical values of the strength of confirmation/rejection to the features.

Formally, for this case the vectors which enter the pattern recognition part are $v_{pro}^T = [x_{1,pro},...,x_{n,pro},y_{1,pro},...,y_{q,pro}]$ and $v_{con}^T = [x_{1,con},...,x_{n,con},y_{1,con},...,y_{m,con}]$, where $x_{i,pro}$ and $x_{i,con}$ are the appropriately transmitted initial features and $y_{i,pro}$ and $y_{i,con}$ stand for the additional features (relevant combinations of the initial ones). Note that the number of combinations concerning PRO- and CON- evidence is not the same, which expresses the fact that some syndromes are used to indicate only presence (or absence) of a disease without any connection to the alternative decision.

(ii) PRO- and CON- evidence are obtained in the form of vectors containing the strengths of the respective features or combinations of features $v_{pro}^T = [\mu_{1,pro},...,\mu_{n,pro},v_{1,pro},...,v_{q,pro}]$ and $v_{con}^T = [\mu_{1,con}, ... , \mu_{n,con},v_{1,con},...,v_{m,con}]$, where $\mu_{i,pro}.$ and $\mu_{i,con} \in [0,1]$ are the confirmation and rejection strength, respectively, for the value of the i-th feature from patient's data, and $v_{i,pro}.$ and $v_{i,con} \in [0,1]$ are the strengths of the additional features.

(iii) μ_{pro} and μ_{con} are completely formed by the AI preprocessing part and enter the pattern recognition part.

In most AI systems the inference engine is supplied with a mechanism to calculate the degree of belief (the confirmation strength) of a single feature value, or of the final decision. Due to the nature of medical reasoning the inferred degree of belief is not equal to 1 - (degree-of-unbelief). This fact has led to the need of inference mechanisms for parallel computation of μ_{pro} and μ_{con} (see e.g., [6]).

4 Pattern Recognition Part

The structure and functioning of this part are determined by the type of its input. For the case (i) the pattern recognition part should implement the mapping

$$\psi : R^{n+q} \times R^{n+m} \longrightarrow [0,1] ,$$

i.e., $\psi(v_{pro},v_{con})$ equals the final degree of confirmation of the hypothesis. This can be rewritten in more detail according to the statement described in Section 2:

$$\psi_1 : R^{n+q} \longrightarrow [0,1] , \quad i.e., \quad \psi_1(v_{pro}) = \mu_{pro}$$

$$\psi_2 : R^{n+m} \longrightarrow [0,1] , \quad i.e., \quad \psi_2(v_{con}) = \mu_{con}$$

and

$$f : [0,1] \times [0,1] \longrightarrow [0,1] ,$$

i.e., $f(\mu_{pro}, \mu_{con})$ equals the degree of confirmation of the final hypothesis.

All mappings ψ_1, ψ_2 and f can take any appropriate nonlinear form. Note that the nonlinearity of f is implied by the data in Table 1. The structure of the pattern recognition part of the scheme for this case is shown in fig 2.

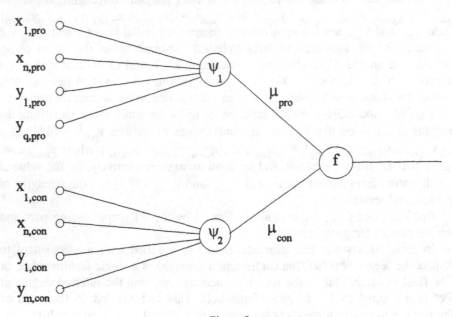

Figure 2

The scheme looks similar to a part of a neural network but in fact it is quite different. All nodes of the scheme are interpretable because their outputs are charged with semantic load. Moreover, the "neurons" here may implement nonlinear functions which may be also non-analytical and on the other hand the tuning parameters of the mappings are not necessarily the weights of the inputs. Therefore backpropagation training may appear inapplicable. Instead, simulated annealing may be used, following the analogy with neural networks. For the sake of clarity and interpretability ψ_1 and ψ_2 may be chosen linear.

Nearly the same is the structure for the case (ii). The difference is in its inputs. Then the mappings ψ_1 and ψ_2 become

$$\psi_1, \psi_2 : [0,1] \longrightarrow [0,1] .$$

Since they operate on membership values different aggregation techniques may be applied from the wide palette of fuzzy decision making: pessimistic or optimistic aggregation, formulas with parametrized degree of optimism, Yager's bag principle, etc. The same considerations about the training of the scheme hold also here.

In the case (iii) the scheme is confined only to aggregation of μ_{pro} and μ_{con}.

The question ensuing is the form of f. It can be seen that the following function coincides with the boundary values from table 1 and implements a smooth transition between them:

$$f(\mu_{pro}, \mu_{con}) = \mu_{pro}^{\alpha} (1 - \mu_{con}^{\beta})$$

α and β are tuning parameters whose values are obtained during the training session.

5 A Potential Application to CADIAG-2

As a potential application field the inference engine of the expert system for internal medicine CADIAG-2 [7,8] can be considered. This comment sketches the conditions which make the application of the combination scheme reasonable.

The fuzzy reasoning model implemented in CADIAG-2 uses two values relating each symptom $X_i \in X$ to the respective disease $D_j \in D$: frequency of occurrence μ^o and strength of confirmation μ^c. On this basis, a diagnostic support is provided inferred by composition of fuzzy relations. Three types of diagnoses are distinguished: confirmed, excluded, and possible. In terms of pro- and con- evidence this decision is as follows:

- confirmed diagnoses $\Leftrightarrow \mu_{pro} = 1.0$
- excluded diagnoses $\Leftrightarrow \mu_{con} = 1.0$
- possible diagnoses $\Leftrightarrow \mu_{pro} \in [\varepsilon, 0.99]$

where ε is a preliminary defined threshold.

In fact μ_{con} is used only when its value is 1.0. Due to the underlying rules, if the negative strength which is computed differs from 1.0 it is not interpreted as a degree of negative evidence. Anyway, the formulation of μ_{con} using the current knowledge base of CADIAG-2 is possible. It can be based on the degree of presence/absence of respective pathognomonic features in the patient data.

Two comments seem necessary:

1.	The decision suggestion for or against a particular hypothesis is based on only one of these values. Note that if $\mu_{pro} = 1.0$ the diagnosis will be viewed as totally confirmed without any doubt, independently on the value of μ_{con}. This reasoning holds even more stronger for the possible diagnoses where the strength of the negative evidence may even prevail.

2.	It can be seen that the transition between confirmed and excluded diagnoses on the one hand, and possible ones on the other is too sharp. Value 0.01 for degree of confirmation is critical to put the diagnosis into the pool of possible ones. This value is also sufficient to prevent excluding of a certain diagnosis although the negative evidence is 0.99. Taking into account the fact that in the formulation of the frequency of occurrence and the degree of confirmation a great deal of subjective assessments are involved [9], a smooth transition appeared more appropriate.

Applying the combination scheme (iii) to this problem may lead to simultaneous accounting for positive and negative strengths on the one hand, and to alleviation of the sharp transition, on the other.

6 An Example from Aviation Medicine

An example with real data was carried out in order to demonstrate the enhanced performance of the scheme in comparison with the linear discriminant analysis considered as a classical technique in pattern recognition.

Data for 295 pilots were used from an examination in a centrifuge with an acceleration profile simulating aerial combat maneuvering (SACM). The problem was to predict if a pilot would exhibit cardiac ectopy during the examination. The parameters for each pilot (measured before the examination) formulating the initial feature set were:

- Number of flight hours;
- Age of the pilot;
- Height of the pilot;
- Weight of the pilot;
- Systolic blood pressure;
- Diastolic blood pressure;
- Heart rate immediately before examination.

Two groups were formed: Pilots without cardiac disorders during SACM, and pilots with single extrasystols (Grade 1 by Lown's scale). Data for pilots with higher grades were omitted from the example.

Applying linear discriminant analysis to this problem an accuracy of 64.1 % was obtained using the resubstitution method, and 60.3 %, assessed by the leave-one-out method.

It can be seen that the parameters are highly nonspecific and cannot be used to build up a rule-based paradigm. Only vague guidelines like "If the age is less than, say 25, or above, e.g., 30, it might be an indication for this type of extrasystols" can be obtained from the expert.

The proposed scheme was used in its second variant. Initial features were transmitted through respective membership functions and the two vectors with membership values $v_{pro}^T = [\mu_{1,pro},...,\mu_{n,pro}]$ and $v_{con}^T = [\mu_{1,con}, ... , \mu_{n,con}]$ enter the second part. No additional features were formed. The types of the membership functions were elicited from the expert's statements about the features. The mappings ψ_1 and ψ_2 performed averaging. Simulated annealing algorithm was used to tune the position and the slope of the respective the membership functions and the parameters of the fuzzy neuron. The classification accuracy obtained was 76.6 %.

7 Conclusions

The paper presents an idea about a classification scheme combining tools from the fields of Artificial Intelligence and fuzzy pattern recognition trying to resemble a physician's decision process. This scheme relies on the powerful AI mechanisms to cope with missing values, to formulate relations between symptoms, to handle rare combinations of features and on a fuzzy pattern recognition conclusion paradigm which can be numerically trained. The expected benefit is an increase in the classification accuracy which would reflect in an improvement in patient care.

References

1. Gelsema E.S. (Ed.) Pattern recognition and Artificial Intelligence in medical research and clinical practice, *Methods of Information in Medicine*, **28**, 1989, 63-65.
2. Chandrasekaran B., A. Goel. From numbers to symbols to knowledge structures: Artificial intelligence perspectives on the classification task, *IEEE Transactions on Systems, Man, and Cybernetics*, **18**, 1988, 415-424.
3. Kuncheva L. Fuzzy multi-level classifier for medical applications, *Computers in Biology and Medicine*, **20**, 1990, 421-431.
4. Kissiov V., S. Hadjitodorov, L. Kuncheva. Using key features in pattern classification, *Pattern Recognition Letters*, **11**, 1990, 1-5.
5. Kuncheva L., R. Zlatev, V. Raicheva. A decision making system in aviation medicine, *Proc. MIE'91, in: Lecture Notes in Medical Informatics*, **45**, 1991, 418-422.
6. Umeyama S. The complementary process of fuzzy medical diagnosis and its properties, *Information Sciences*, **38**, 1986, 229-242.
7. Adlassnig K.-P. Fuzzy set theory in medical diagnosis, *IEEE Transactions on Systems, Man, and Cybernetics*, **SMC-16**, 1986, 260-265.
8. Adlassnig K.-P., W. Scheithauer, G. Kolarz. Fuzzy medical diagnosis in a hospital, In: H. Prade and C. Negoita (eds.) Fuzzy Logic in Knowledge Engineering, Verlag TUV Rheinland, Germany, 1986, 275-294.
9. Adlassnig K.-P. Uniform representation of vagueness and imprecision in patient's medical findings using fuzzy sets, in: R. Trappl (ed.) Cybernetics and Systems'88, Kluwer Academic Publishers, 1988, 685-692.

Fuzzy Concepts for Predicting the Behaviour of other Drivers on a Highway

Friedhelm Mündemann*)

FH Brandenburg, O-1800 Brandenburg

Abstract. The task to solve is the automatic "safe" guidance of an autonomous mobile vehicle along a German highway (AUTOBAHN) at moderate to high speed. To assist the necessary decision making process of how to behave "safely" in an actual traffic situation, data about the road itself (e.g. number of lanes, lane width, lane markings), traffic signs and other traffic participants around the own car have to be gathered, represented and interpreted appropriately. In this contribution fuzzy concepts are suggested in the development and use of part of an internal world model, based on which sensor data about other traffic participants can be evaluated and a prediction of their short to medium range future behaviour (i.e. their expected movements) can be derived such that decisions about the own future behaviour can be drawn.

Keywords: Preparation of decisions with uncertain/vague data
                       Real-time fuzzy expert systems
                       Application: autonomous mobile vehicles

1 Introduction

The task to solve is the automatic "safe" guidance of an a̲utonomous m̲obile v̲ehicle (AMV) along a German highway (AUTOBAHN) at moderate to high speed. This environment is more or less standardized. Besides the inevitable engineering problems, lots of information-processing problems have to be solved in real-time: providing appropriate data from the outer world, interpreting them correctly, deriving sound hypotheses from these data, building a correct model of the outer world inside the system and deriving decisions about the AMV's future "safe" behaviour while performing a mission from a point A on the AUTOBAHN to a point B on the AUTOBAHN according to a driver's wishes.

In this contribution the development and use of part of an internal world model is shown, based on which sensor data about other t̲raffic p̲articipants (TP) can be evaluated and a prediction of their short to medium range future behaviour (i.e. their expected movements) can be derived such that decisions about the own future behaviour can be drawn. For the following sections it is assumed that the reader is familiar with basic knowledge about fuzzy data processing as for example is described in /KAND 91, KRUS 87, LEE 90, MAYE 93, ZADE 65, ZADE 89/.

The world modelled is the German AUTOBAHN with multiple lanes and several classes of TP e.g. cars, busses, lorries or trucks (see fig. 1). Data from the outside world are sampled using video cameras.

Figure 1: Typical scene from an AUTOBAHN

---

*) The research was performed while the author was at the Universität der Bundeswehr München, 8014 Neubiberg

Among the problems to solve while developing an autonomous mobile vehicle, three problems shall be pointed out.

First, when using off-the-shelf cameras as sensors for data acquisition from the outer world, the nearly unavoidable noise degrades the quality of picture data and the rather low resolution tends to hide important details of the scene under sometimes just one pixel.

Second, as there is usually no explicit communication between the own car and other traffic participants (TP), one has no knowledge about their intentions while moving along the AUTOBAHN. Instead, one has to decide upon one's own behaviour only based on visual data from other TP, aggregated over some time in the past.

Third, human knowledge about driving cars "safely" is usually not taught by showing people how to solve the appropriate differential equations, but by teaching them a lot of rules of thumb instead.

Therefore one has to deal with uncertain, noisy sensor data while being faced with the task to judge about the possible future behaviour of moving hypothetical bodies around the own car (i.e. the AMV) that are (mostly) guided by humans.

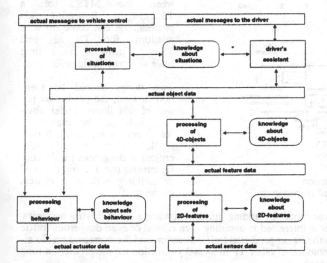

Figure 2: System architecture for an autonomous mobile vehicle

At the Universität der Bundeswehr München researchers from the Aerospace Department successfully developed the basic theoretic and practically usable methods for vision-based automatic driving /DICK 86, DICK 87, DICK 91, GRAE 92/.

In the sequel an interdisciplinary cooperation between the Aerospace Department and the Computer Science department led to the development of a layered architecture for an autonomous mobile vehicle /MÜND 91, NIEG 93/, which consists of a sensor data processing layer, a hypothesis generation layer for establishing a world's model of the AUTOBAHN, a decision making layer and a driver interface, incorporating a mission planning module (fig. 2).

An especially equipped van (DB 508) is operationable as a testbed. Due to the very successful work of the Aerospace Department collegues all necessary engineering solutions are provided.

To assist the necessary decision making process of how to behave "safely" in an actual traffic situation, data about the road itself (e.g. number of lanes, lane width, lane markings), traffic signs and other traffic participants around the own car have to be gathered, represented and interpreted appropriately. The Kalman filter used by the Aerospace Department is an optimal estimator for movement prediction on a short time scale, typically at the system cycle of 40-80 milliseconds. At longer time scales, estimation of movements of other TP is expected to support the decision making process as it can extract typical movement trends that may lead to "traffic patterns" the reaction to can possibly correct, prevent or evoke decisions that are otherwise based on data derived on a shorter time scale alone.

The remainder of the paper is organized as follows. In section 2 the procedure for predicting the behaviour of other drivers is described. In section 3 the evaluation of a simple traffic situation is presented as an example. A conclusion summarizes the state of the work.

2 Procedure for predicting the behaviour of other drivers

The main idea of predicting the future behaviour of other TP on a longer time scale is as follows. At each time step the actually perceived traffic situation (TS$_{now}$) consists of a snapshot of the TP-configuration around the AMV. Each TP is described by a measurement vector consisting of the distance in x-direction relative to an "observing" car and the relative deviation in y-direction, the relative velocities in x- and y-direction and data about the vehicle's body (length, width, height), namely <distance, deviation, velocity$_x$, velocity$_y$, length, width, height>. Based on the set of measurement vectors, one can derive a set of static "relations" between all members of the perceived traffic situation (MTS).

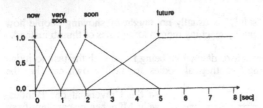

Figure 3: Fuzzy time scale

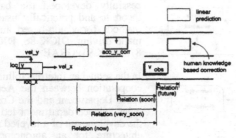

Figure 4: Sketch of the linear prediction model with human knowledge based corrections

Assuming that for short time intervals the consequences of physical laws dominate the drivers behaviour decisions that may be based on their own will, velocities are assumed to be constant for the prediction interval. This leads to a set of relations between the MTS for a TS$_{very\ soon}$. Repeating this process, one can come up with relations for TS$_{soon}$ und TS$_{future}$ using a fuzzy time scale as shown in figure 3.

Two things should be mentioned: First, the prediction power of this linear model obviously is limited, but may serve for situations with smooth traffic flow. Second, it may predict critical or dangerous traffic configurations (i.e. leaving the road or colliding with other objects on the road).

Therefore one can correct this linear model by adding human knowledge about driving assuming that (hopeful-ly) very human driver is interested in avoiding such critical or even dangerous traffic situations. The difference between the linear predicted traffic situation and the corrected traffic situation reflects the expected accelerations of other TP in x- and y-direction for the prediction time interval. Figure 4 sketches the approach.

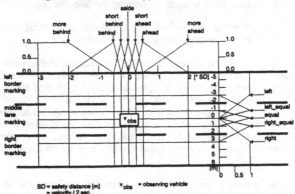

Figure 5: Fuzzy geometric model of a part of the AUTOBAHN

The procedure of predicting the future behaviour of other TP consists of several design steps /HEIN 92/.

First, one has to decide upon the universe of values to be fuzzified and defuzzified, i.e. the measurement vector data space. The resolution must be fine enough to be able to react on small changes in the data if necessary and coarse enough to smoothen the noisy sensor data and to be tractable during real-time computations (see fig. 5).

Figure 5 shows the established fuzzy geometric model of that part of the AUTOBAHN that currently surrounds the own car. It is an observer centered model and covers a three-lane road. This model can be used as an internal representation of the outer world and may serve for keeping track of hypothetical moving bodies and for re-identifying them as relevant TP /MÜND 92/.

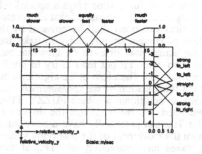

Figure 6: Linguistic variables and
fuzzy sets for relative velocities

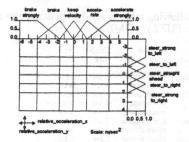

Figure 7: Linguistic variables and fuzzy sets
for predicted relative accelerations

Figure 6 shows the linguistic variables and their associated fuzzy sets for the relative velocities in x- and y-direction.

In order to establish a common universe of discourse, the set of possible relations between MTS has to be defined. Such a relation is defined by labelling combinations of relative locations and velocities between the observing vehicle A and another TP B using human understandable terms. These relations correspond to typical binary traffic patterns, e.g. A drives left ahead of B, A is rammed by B from behind, A passes B on the left or A follows B.

We use 15 relations between the observing vehicle and other TP that are in front of it, 18 relations for TP aside and 13 relations for TP behind the observing vehicle.

To come up with a prediction of accelerations in x- and y-direction, the linguistic variables and associated fuzzy sets as shown in figure 7 are used.

Second, one has to derive fuzzy rules to model the possible movement behaviour of a TP together with its consequences (i.e. for the possible development of the actual traffic situation).

The necessary knowledge acquisition has been done by case studies and interviews, based on the question "How would you expect the other TP to react in this situation?". The resulting rule set is represented using look-up tables, one for each TP class. This rule set has to reflect among other things the different vehicle dynamics of the members of distinct TP classes, road geometry and common federal rules for movements of TP on AUTOBAHNEN, e.g. traffic signs or lane markings.

Third, one has to evaluate the actual world model every system cycle according to the rule set to keep track with the high dynamics in movements of other TP and to be able to react on them as soon as possible. Figure 8 shows a dataflow diagram of the whole algorithm.

In a first step (R1) the x- and y-values of the measurement vector are fuzzified and grouped together to model the location relative to the "observing" car, the same is done with the relative velocities (R2). The TP's outside measures are fuzzified to be used for the determination of the TP class (R3). In the next step (R4), a location for the TP is predicted, assuming that its last velocity components are kept in the near future. The results from R1 and R2 and the results from R4 in the steps R5 and R6 are used to determine "binary" traffic situation patterns, that are looked up in a table, the behavioural consequences of which have been determined in the knowledge acquisition step described earlier. Then, in the last two steps, the derived expected behaviour is determined (R7), defuzzified (R8) and forms the acceleration vector that seems to be most probable for the traffic situation found wrt one "observing" car.

Combining all prediction results from all "observing" cars take into account all expected "influences" upon the future possible moving behaviours of all TP found in the actual traffic situation and yield as the final result the most probable expected moving behaviour of all cars wrt the AMV. As such, the result of the predictions is a possible world state at a future point in time or at several future points in time.

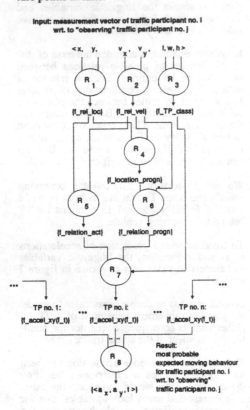

The concept of fuzzy time plays a special role in the algorithm. It is assumed that all potentially critical situations shall be covered by the prediction. Human drivers not only take into account actual risks or even dangers that may come from other TP but tend to try to predict future risks, too. Therefore in the algorithm during the fuzzification of the measurement vector data an additional attribute FUZZY_TIME is generated. It's value reflects the time to go until a TP_n can potentially have a risky influence on the movement of the observing vehicle that makes an observation wrt TP_n at time t_{now}. For example, a TP_n that is observed at a location *more behind* and that is *much faster* can potentially have a risky influence at a *future* point in time.

The following table is used to generate the attribute FUZZY_ TIME.

Relative location(x)	Relative velocity(x)	Crisp time	Value of FUZZY_TIME
more behind	much faster	5 sec	future
behind	much faster	3 sec	soon
short behind	much faster	1 sec	very soon
aside	much faster	1 sec	very soon
aside	much slower	1 sec	very soon
short ahead	much slower	1 sec	very soon
ahead	much slower	3 sec	soon
more ahead	much slower	5 sec	future

Figure 8: Dataflow diagram of the algorithm for behaviour prediction

3 Evaluation of a simple traffic situation

Three remarks in advance of the example: First, the maximum operator is used throughout the model. Second, the value for FUZZY_TIME that is generated during the fuzzification step for the fuzzy location vector, is propagated throughout the algorithm. Third, the tables that contain human knowledge about driving are not densely filled. The reason is that during the acquisition process it became clear that not all possible combinations that could form a relation will contribute to "useful" predictions. Consider for example a TP that is more ahead and equal in deviation and that is much faster straight. Human experience tells that this TP will be out of sight soon. Table entries that are not used are marked *omit*.

It also depends on the personal preferences of the drivers that are asked during the acquisition steps which table entries he or she is considering to be not useful in predicting critical situations.

Consider the following example of a simple traffic situation as shown in figure 9. There are two vehicles (TP_1, TP_2) on a two-lane road.

The two TP are described by the following measurement vectors at t_{now}:

TP₁:
    <distance   = 0.0,
    deviation = 0.0,
    velocity$_x$ = 0.0,
    velocity$_y$ = 0.0,
    length   = 4.0,
    width   = 1.6,
    height   = 1.5>

TP₂:
    <distance   =-30.0,
    deviation = 0.0,
    velocity$_x$ = 20.0,
    velocity$_y$ = -0.5,
    length   = 4.0,
    width   = 1.6,
    height   = 1.5>

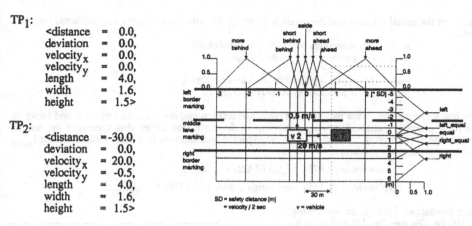

Figure 9: Sketch of an example of a traffic situation.

The prediction steps work as follows. The relative distance of 30 m at a velocity of 20 m/sec (TP₂) is recalculated to 0.778 * safety_distance, which is assumed to be 2 seconds. This corresponds to an actual safety distance of 36 m. The fuzzification step for the relative distance comes up with the vector:

    <very soon: -, soon: ahead (0.75), future: more ahead (0.125)>

As there is no location responding with a degree > 0.0, the entry for *very soon* is empty. Fuzzification of the relative deviation results in *<very soon: equal (1.0)>*. Both data are grouped to the following vector LOC:

    <very soon: -, soon: ahead equal (0.75), future: more ahead equal (0.125)>.

Fuzzification of the relative velocities yields the vector:

    <very soon: much slower (0.875)> for velocity_x and
    <very soon: strong_to_right (0.125), to_right (1,0), straight (0.5)> for velocity_y

that are combined to the vector VEL:

    <very soon: much slower strong_to_right (0.125), much slower to_right (0.875), much slower straight (0.5) >.

The data from LOC and VEL are used to look up the actual relation at t$_{now}$ wrt t$_{soon}$ and t$_{future}$ using a transition table: location × velocity → location. The results are for
t$_{soon}$:
much slower strong_to_right (0.125) and *ahead equal (0.75)* becomes *omit*.
much slower to_right (0.875) and *ahead equal (0.75)* becomes *TP1 strongly oppresses TP2 from ahead (0.75)*.
much slower straight (0.5) and *ahead equal (0.75)* becomes *TP1 strongly oppresses TP2 from ahead (0.5)*.
t$_{future}$:
much slower strong_to_right (0.125) and *more ahead equal (0.125)* becomes *omit*.
much slower to_right (0.875) and *more ahead equal (0.125)* becomes *TP1 is passed on the left by TP2 (0.125)*.
much slower straight (0.5) and *more ahead equal (0.125)* becomes *TP1 slightly oppresses TP2 from ahead (0.125)*.

The inference step for the predicted location uses the same data from LOC and VEL and results in
t$_{soon}$:
    much slower strong_to_right (0.125) and *ahead equal (0.75)* becomes *omit*.
    much slower to_right (0.875) and *ahead equal (0.75)* becomes *aside right equal (0.75)*.
    much slower straight (0.5) and *ahead equal (0.75)* becomes *aside equal (0.5)*.
t$_{future}$:
    much slower strong_to_right (0.125) and *more ahead equal (0.125)* becomes *omit*.
    much slower to_right (0.875) and *more ahead equal (0.125)* becomes *aside right (0.125)*.
    much slower straight (0.5) and *more ahead equal (0.125)* becomes *aside equal (0.125)*.

Based on the actual relation and the predicted location the inference step for the predicted relations yields:

> soon: *TP₁ strongly oppresses TP₂ from ahead (0.500).*
> soon: *TP₁ gives way to TP₂ to behind right (0.750).*
>
> future: *TP₁ slightly oppresses TP₂ from ahead (0.125).*
> future: *TP₁ is passed on the left by TP₂ (0.125).*

The last inference step takes into account the actual relation and the predicted relation and looks up the most probable expected moving behaviour in the correction tables that represent the human knowledge about driving to come up with the expected accelerations in x- and y-direction for t_{soon} and t_{future}. This step predicts

$$< brake\ (0.875),\ to\_left\ (0.625)> \text{ for } t_{soon}$$

$$< keep\ velocity\ (0.125),\ steer\ straight\ ahead\ (0.125)> \text{ for } t_{future}.$$

Both predictions indicate an overtaking of TP_1 by TP_2 (see fig. 10) and seem to make sense in the example although one would not expect the "brake"-prediction. But this is clear if one looks at the table entries: they have been expressed following the rule "priority to braking versus steering in order to escape critical situations".

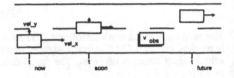

Figure 10: Qualitative plot of the predicted situation development

Figure 11 shows a more complicated example that as the final prediction step results in the estimation that TP no. 1 and 2 continue their last shown behaviour, TP no. 3 is expected to brake, TP no. 4 is just moving to the left and TP no. 5 is expected to change lane to the right.

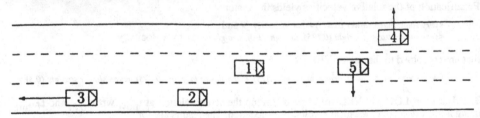

Figure 11: Example of a predicted traffic situation on a three-lane AUTOBAHN

As an advantage of using fuzzy concepts one can easily derive an explanation module from the fuzzy inference process which shows the actual environment graphically and the predicted behaviours either textually or graphically.

For making final decisions upon the own next behaviour, in our integrated approach we further combine the information from the fuzzy-based world model with numerical calculations of possible vehicle trajectories of the own car and of other TP to ensure that at every point in time a certain safety envelope, the size of which depends on the velocity of the own car, is free from other TP /MÜND 92/. Thus, fuzzy and numerical calculations are used to assist each other to come up with robust data for decision making within the AMV.

Conclusion

In this contribution a first version of an algorithm is presented that takes measurement data about moving bodies around an observing vehicle that are derived from hypothesis generation steps to generate movement predictions on a longer time scale (typically one to several seconds) to assist the decision making process for "safely" guiding an autonomous mobile vehicle along German highways at moderate to high speeds.

It should be kept in mind, that there are lots of parameters to be tuned in the algorithm:

First, there are static design decisions to be optimally taken e.g. the definition of the linguistic variables, the resolution, shape and overlapping of their associated fuzzy sets. For the algorithm presented the design decisions can be seen in the figures 3, 5, 6, 7.

Second, there is the set of traffic patterns to be defined to come to appropriate relations.

Third, there is the knowledge acquisition step, to be done for each class of TP seperately. In our algorithm, we use one table for vehicles, one for lorries and one for vans and busses. It seems achievable that the algorithm may be adapted to several different styles of driving e.g. conservative and non-risky or more sportive. The table entries may have to be changed accordingly.

Fourth, there is the right choice of the fuzzy operators for the inference and defuzzification steps. In the algorithm described, the maximum operator and the COA (center of area) operator are used.

Fifth, the propagation of the values for the attribute FUZZY_TIME throughout the algorithm can be elaborated to result in the prediction of a trajectory that is supposed to be taken by the other TP.

As a result, the "calibration" of the algorithm has to be done carefully.

The programmed prediction system consists of a table editor for managing the fuzzification tables and the rule sets, a separate run-time module for calculating the behaviour predictions and an object-oriented simulation environment.

The module has been programmed in C and tested using simulation data from real traffic scenes with promising results. The algorithm is $o(n^2)$, so the execution time is depending on the number n of TP actually covered by the world model. It executes on a 33Mhz-80486 PC within a cycle time of 80 milliseconds for up to 7 TP /HEIN 92/. The module is intended to continously run at that cycle time in the test-van.

The module shall be integrated in a multi-transputer environment next and evaluated thereafter on short-distance runs on our test-highway in simple traffic situations and on long-distance runs on the AUTOBAHN by the end of this year.

References

/DICK 86/ Dickmanns E.D.:
Computer Vision in Road Vehicles - Chances and Problems, ICTS-Symposium on "Human Factors Technology for Next-Generation Transportation Vehicles", Amalfi, Italy, June 16-20, 1986.

/DICK 87/ Dickmanns E.D.:
Object Recognition and Real-Time Relative State Estimation Under Egomotion, Paper presented at the NATO Advanced Research Workshop on "Real-Time Object Recognition...", Maratea, Italy, Aug. 31-Sept.4, 1987.

/DICK 91/ Dickmanns, E.D.:
　　　　4D dynamic vision for intelligent motion control. In: C. Harris (Ed.): Special issue
　　　　of the Int. Journal for Engineering Aplications of AI (IJEAAI) on 'Intelligent Auto-
　　　　nomous Vehicle Research', 1991.

/GRAE 92/ Graefe, V., Kuhnert, K.D.:
　　　　Vision-based autonomous road vehicles. In: I.Masaki (Ed.): Vision-based vehicle
　　　　guidance, Springer-Verlag, Berlin, 1992, p. 1-29.

/HEIN 92/ Heinz, S., Richter, D.:
　　　　Entwicklung eines echtzeitfähigen Prognosemoduls für das Verhalten von Verkehrs-
　　　　teilnehmern mit Hilfe von Methoden der Fuzzy-Logik, Diplomarbeit, Universität der
　　　　Bundeswehr München, Fachbereich Informatik, Neubiberg, September 1992.

/KAND 91/ Kandel, A.(ed.):
　　　　Fuzzy Expert Systems, CRC Press, Boca Raton, FL/USA, 1991.

/KRUS 87/ Kruse, R., Meyer, K.D.:
　　　　Statistics with Vague Data, D. Reidel Publ. Comp., Dordrecht, 1987.

/LEE 90/ Lee, C.C.:
　　　　Fuzzy Logic in Control Systems: Fuzzy Logic Controller, IEEE Transactions on Sy-
　　　　stems, Man, and Cybernetics, Vol. 20; No.2, March/April 1990, p.404-435.

/MAYE 93/ Mayer, A., Mechler, B., Schlindwein, A., Wolke, R.:
　　　　Fuzzy Logic, - Einführung und Leitfaden zur praktischen Anwendung, Addison-
　　　　Wesley Publ. Comp., Bonn, Paris, Reading/Mass., 1993.

/MÜND 91/ Mündemann, F.:
　　　　A Uniform Layered Software Architecture for Knowledge Processing of Higher
　　　　System Levels within the AUTOBAHN-AUTOPILOT, 3rd PROMETHEUS
　　　　Workshop, Compiegne, October 1991.

/MÜND 92/ Mündemann, F.:
　　　　Checking the safety envelope, - a prediction based concept for collision avoidance,
　　　　3rd PROMETHEUS workshop on collision avoidance, Nürtingen, June 1992.

/NIEG 93/ Niegel, W.:
　　　　Knowledge processing at higher system levels of autonomous road vehicles, Conf.
　　　　on Autonomous Intelligent Vehicles, Southhampton, April 1993 (to be published).

/ZADE 65/ Zadeh, L.:
　　　　Fuzzy Sets, Information and Control, 1965.

/ZADE 89/ Zadeh, L.:
　　　　Knowledge Representation in Fuzzy Logic, IEEE Transactions on Knowledge and
　　　　Data Engineering, Vol. 1, No. 1, March 1989, p.89-100.

Design of a Fuzzy Car Distance Controller

Alexandra Weidmann

Department of Computer Science, FU Berlin
D-1000 Berlin 33

Abstract. The paper presents a model of a fuzzy car distance controller. In contrast to conventional distance controllers this model additionally takes into account fuzzy information about the driver and about the environment. First the principle of the sharp car distance controller and its deficits are explained. The analysis shows that fuzzy models of the driver and of the environment are necessary to improve the car distance controller. A realisation of the fuzzy model is presented. Some simulation results for the fuzzy distance controller are demonstrated.

1 Introduction to car distance control

A car distance controller is responsible for maintaining a minimal distance between two cars driving on the same lane in the same direction, so that the car driving behind is guaranteed to stop safely in an emergency situation. The controller can also be used for the optimisation of car distances to maximise the road capacity [Leutzebach 1979]. The controller calculates the emergency distance and changes the velocity based on the difference between the desired and actual distance to the car in front.

Figure 1 demonstrates an example of such a scenario, with cars C1 and C2. Car C2 is equipped with sensors (e.g. radar) measuring the actual distance of car C2 to car C1 and their relative speed $v_1 - v_2$. Speed (v_1) of car C1 can be calculated by measuring the actual distance d_a twice in order to determine the relative speed $v_1 - v_2$.

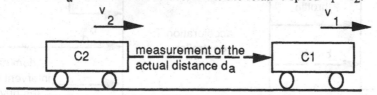

Fig. 1. Measurement of the actual distance d_a between two cars C1 and C2

The desired distance d_d results from the velocities v_1 and v_2, brake retardations a_1 and a_2 of cars C1 and C2 respectively and from the reaction time r of the driver of car C2. The desired distance d_d is given by the difference of the braking distances of C1 and C2 plus the average reaction distance of the car C2.

$$d_d = \frac{v_2^2}{2 \cdot a_2} - \frac{v_1^2}{2 \cdot a_1} + r \cdot v_2 \qquad (1)$$

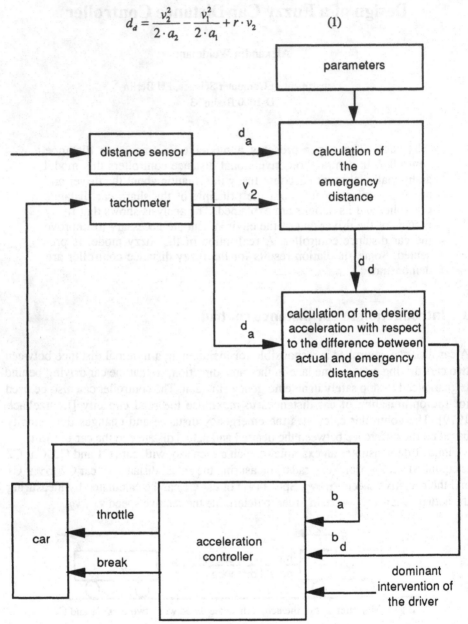

Fig. 2. Concept of a sharp distance controller

A sharp distance controller (figure 2 [Wocher 1971]) first calculates the desired emergency distance using the sensor values and parameters r, a_1, and a_2 The desired acceleration b_d is determined by the desired distance d_d, the actual distance d_a, and additional optimisation criteria. Such an optimisation criteria might be to maximise the road capacity by minimising the difference between actual and desired emergency distance. The acceleration controller changes the velocity of the car C2.

The parameters r, a_1 and a_2 involved in the calculation of the desired emergency distance are constant and they are not adapted to changes such as changing road conditions. As a result, the desired distance which influences the controller can not always be calculated accurately. This is further elaborated upon in the remainder of this section providing the reasons for the proposal made in this paper.

The reaction time r is the time the driver needs to react to a break signal from the car in front. It includes affection, consciousness, decision time and the time a driver needs to move his foot from the accelerator to the brake pedal and the time until the brake becomes effective. The minimal amount of time is estimated to take 0.4 sec by [Färber 1986]. The average reaction time varies in different studies between 0.4 sec [Färber 1986], 0.6 sec [Hör 1983] and 0.8 sec [Burkhardt 1976].

The brake retardations a_1 and a_2 depend on the conditions of the breaks, of the tyres, and of the road. The highest brake retardations for an emergency brake a_1 and a_2 are fixed by 8 m/s^2 and 5 m/s^2 respectively for a normal, dry road [Wocher 1971]. For safety reasons a_2 is always set to be smaller than a_1. In experiments the majority of test drivers accepted a value of 6 m/s^2 for a_2 [Färber 1986].

The calculation of the desired distance normally uses fixed values for the reaction time r of the driver in car C2 and for the brake retardations a_1 and a_2. Equation 1 only delivers the desired distance d_d for an emergency brake under optimal conditions.

Exact measurements of brake retardations taking into account road conditions are not feasible due to the lack of appropriate sensors. Additionally the actual evaluation of a driver's reaction time is very costly. Those factors have to be measured in changing situations such that a maximum agreement between the calculated and the real necessary emergency distance is reached. Statistics of car accidents show that an actual distance calculated with the two second count method is not always good enough in order to avoid rear-end collisions [Nicklisch 1989].

2 Improvement of the sharp distance controller

Regarding the structure of the sharp distance controller (figure 2), it is obvious that it will suffice to influence the calculation of the desired distance to improve the whole control. The rest of the controller remains unchanged.

To determine the reaction time r, it is possible to measure the time between the lightening of the stop lights of the car ahead and the activation of the break. The additional technical effort (e.g. optical sensors) is considerable. The determination of brake retardations a_1 and a_2 is even more difficult because they depend on many complex state variables.

In this paper an alternative method is proposed to overcome the difficult task of measuring the reaction time r and the brake retardations a_1 and a_2. These parameters can be fixed to standard values. A fuzzy rule base determines their variations due to real world conditions. To improve the distance controller additional information has to be included properly in the emergency distance calculation and it must be possible to find the information with a reasonable technical effort.

The solution proposed in this paper is to model the variation of the reaction time using fuzzy information about the driver. In the same way the brake retardations a_1 and a_2 can be changed by suitable fuzzy information. Both factors influence the value of the desired distance d_d thus reducing the risk of rear-end collisions.

The reaction time depends on many complex factors like illness, medication, day time etc. Not all influences are known and the interactions between the influences are not well understood [Herberg 1985]. Although this information is not available or incomplete, it is still possible to make inferences about the emergency distance from an incomplete, coarse, and inexact model of the driver and the environment using fuzzy logic. The next section provides the details of a fuzzy distance controller proposed in this paper.

3 Design of the fuzzy distance controller

A fuzzy controller yields better results than a conventional controller if exact data for the calculation of the real necessary emergency distance are not available. Without going into details it should be noted that principal concepts of one part of the model developed in this section are based on the results of traffic psychology ([Sömen 1984], [Näätänen 1976], [Wilde 1978]).

Two different models have been designed, the driver model and the environment model (figure 3). The driver model estimates the risk caused by the variation of the reaction time of the driver. The following influencing factors have been analysed: fitness, fatigue, experience, view, age, and driving style. The factors are judged by their risk potential. Each factor is represented by a linguistic variable. The attributes of a factor (terms) are plotted against the value of the corresponding linguistic variable (figure 4). The linguistic variable f_risk represents the result inferred from all the factors mapped to linguistic variables.

The environment model takes into account the factors road surface (good asphalt, normal asphalt, cobblestone pavement, crushed stone, and sand) and the condition of the road surface (dry, humid, wet, snow-covered, icy). From those the danger foreseen in the influences of the environment is derived. These influences and their attributes are described as linguistic variables and terms respectively as in the driver model.

Correlations are assumed between fatigue and view additionally between fatigue and the condition of the road surface. Other correlations are ignored. Which correlations really exist and how they influence each other are still research topics. The fuzzy estimation contains two components, the risk and the danger estimation. With the result of the risk estimation, the reaction distance is influenced by changing of the reaction time r. With the result of the danger estimation, the stopping distances are influenced by changing the brake retardations a_1 and a_2.

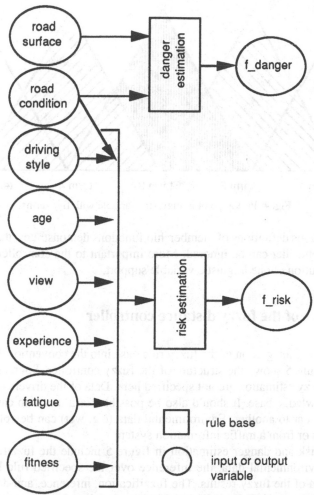

Fig. 3. Model of the risk and danger estimation

All linguistic variables are constructed in the same way (figure 4). Experiments during the tuning of the membership functions have shown that variations of the shape of the membership functions (triangular and trapezoidal) have no significant influence of the results. Such effects have also been observed in [Gupta 1988] and [Hetzheim 1991]. If triangular and trapezoidal functions are defined in a normalised interval of length one, they are application independent [Palm 1991]. This applies to those fuzzy sets which can be used for states and control variables. The rules for the risk and danger estimation have only one term in their preconditions or the terms are connected by And or Or. All rules contain only one term in their conclusion.

The attributes of the terms which represent the various influences are ordered by growing risk or danger potential. The evaluation of the rule base yields one result for the f_risk and one for the f_danger. The results of their defuzzification are included in the calculation of the sharp desired distance.

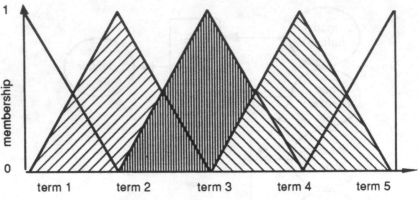

Fig. 4. Prototype of a linguistic variable with five terms

The test of various definitions of membership functions demonstrates, that their influence on the controller can be ignored. More important to the controller is the order and the distribution of one linguistic variable support.

4 Structure of the fuzzy distance controller

In this section the integration of the fuzzy rule base into the conventional controller is presented. Figure 5 shows the structure of the fuzzy controller. Sources of information for the fuzzy estimation are not specified here. Data of the drivers can be memorised in a knowledge base. It should also be possible for a driver to carry his or her data from one car to another. Environmental data (e.g. wet) can be collected by car owned sensors or from a traffic information system.

The fuzzy risk and danger estimation in figure 5 include the fuzzification of the driver and environmental data, the inference over the specified rule base, and the defuzzification of the fuzzy results. The fuzzification, inference, and defuzzification mechanism of the tool TILShell [Togai 1990] are applied for the estimation. To use the fuzzification mechanism the data have to be normalised to the interval 0 .. 255 [Weidmann 1992]. Given the fuzzified data, the rule base can now infer f_risk and f_danger. The defuzzified results are called risk and danger. Taking into account the factors risk and danger equation 1 is refined to yield the modified emergency distance. How this is performed is shown in the next chapter and results in equation 2. The rest of the controller remains unchanged.

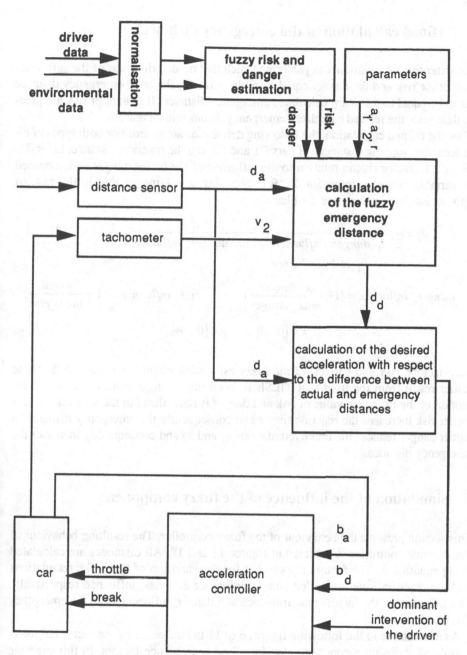

Fig. 5. Prototype of the fuzzy distance controller

5 Refined calculation of the emergency distance

The extension of equation 1 is performed such that no deterioration of the safety occurs. If the risk and the danger equal zero, the refined calculated emergency distance has to be equal to the sharp calculated emergency distance. If danger or risk are greater than zero, the refined calculated emergency distance is increased.

For the refined calculation, the following definitions are given: For both types of distances, the stopping distances of cars C1 and C2 and the reaction distance, the influence of the factor risk in relation to the influence of the factor danger is determined by variables x and y (equation 2). E.g. choosing y = 1 means that in the case of highest risk the reaction time doubles.

$$d_d = \frac{v_2^2}{2 \cdot a_2 \cdot danger\_influence} - \frac{v_1^2}{2 \cdot a_1 \cdot danger\_influence} \qquad (2)$$
$$+ r \cdot v_2 \cdot risk\_influence$$

$$danger\_influence = (1 - \frac{x \cdot danger}{max\_danger}) \qquad risk\_influence = (1 + \frac{y \cdot risk}{max\_risk})$$

$$x \in [0...1[\qquad y \in [0...\infty[$$

Both factors, risk and danger, of the fuzzy estimation range from 0 and 255. These values result from the used tool TILShell. With max_danger = max_risk = 255 the support of the influence value of risk and danger is normalised to the interval 0 .. 1. A higher risk increases the reaction time r and consequently the emergency distance. A higher danger reduces the brake retardations a_1 and a_2 and consequently increases the emergency distance.

6 Simulation of the influence of the fuzzy component

This section presents the behaviour of the fuzzy controller. The resulting behaviour of the distance controller is depicted in figures 11 and 18. All distances are calculated using equation 2. The factors responsible for the variation of the brake retardations and the reaction time are called danger_influence and risk_influence respectively. The influence of the factors risk_influence and danger_influence onto the emergency distance is shown.

As an example in the following figures 6 to 11 the influence of the factor fatigue is simulated choosing a constant value for all other influence factors. In this example the fixed values are: driver's age = 30, day time = 8 a.m., view = very_good, driving style = defensive, road surface = asphalt, the condition of the road surface = dry, speed $v_1 = v_2 = 100$ km/h, the reaction time r = 0.5 sec, the brake retardation $a_1 = 8$ m/s^2, and the brake retardation $a_2 = 6$ m/s^2.

As a second example the influence of the road condition is simulated. The result is shown in figure 12 to 18. In this example the fixed values are the same as in the previous example, additionally fatigue is fixed to "very fit". All graphs were produced using a special simulation environment including the normalisation of the input values [Weidmann 1992]. The result values of the risk and the danger are inferred by the max-min inference method and the centroid defuzzification [Zimmermann 1991]. For the first example the following rules are relevant:

IF fatigue IS alert	THEN risk = very_low
IF fatigue IS awaken	THEN risk = low
IF fatigue IS tired	THEN risk = high
IF fatigue IS very_tired	THEN risk = very_high
IF fatigue IS very_v_tired	THEN risk = very_v_high
IF fatigue IS tired AND street IS bad	THEN risk = very_high
IF fatigue IS tired AND view IS bad	THEN risk = very_high

For the second example the following rules are relevant:

IF street IS very_good	THEN danger = very_low
IF street IS good	THEN danger = low
IF street IS bad	THEN danger = high
IF street IS very_bad	THEN danger = very_high
IF street IS very_v_bad	THEN danger = very_v_high

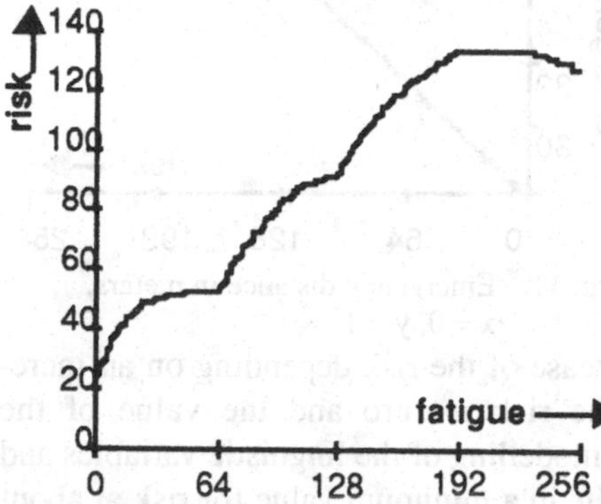

Fig. 6. Risk depending on fatigue

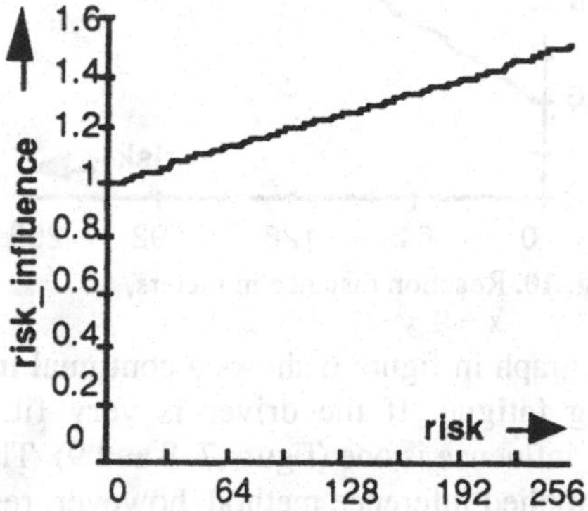

Fig. 7. Risk_influence depending on risk with y = 0.5

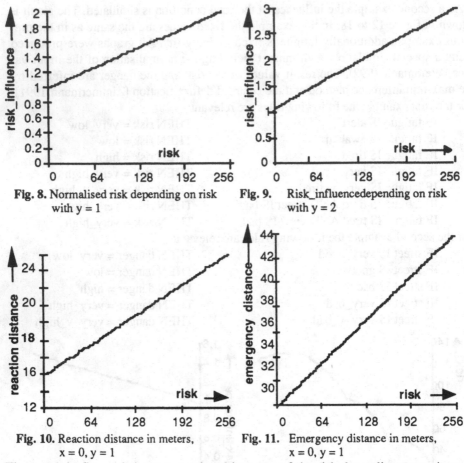

Fig. 8. Normalised risk depending on risk
with y = 1

Fig. 9. Risk_influencedepending on risk
with y = 2

Fig. 10. Reaction distance in meters,
x = 0, y = 1

Fig. 11. Emergency distance in meters,
x = 0, y = 1

The graph in figure 6 shows a continual increase of the risk depending on an increasing fatigue. If the driver is very fit, the risk is zero and the value of the risk_influence is one (figure 7, 8 and 9). The modelling of the linguistic variables and the applied inference method, however, results in a minimum value for risk at about 20 (see figure 6). This minimal value always has to be deducted from the inferred risk value for the calculation of the reaction distance (see figure 10). The same applies for the calculation of the stopping distances. Figure 11 shows the resulting emergency distance.

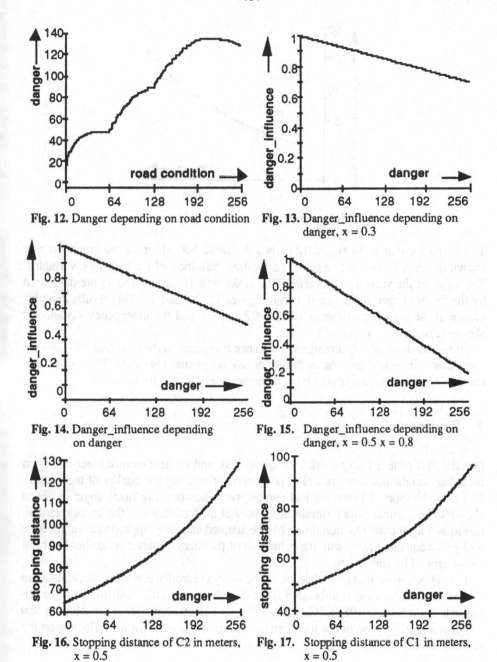

Fig. 12. Danger depending on road condition

Fig. 13. Danger_influence depending on danger, x = 0.3

Fig. 14. Danger_influence depending on danger

Fig. 15. Danger_influence depending on danger, x = 0.5 x = 0.8

Fig. 16. Stopping distance of C2 in meters, x = 0.5

Fig. 17. Stopping distance of C1 in meters, x = 0.5

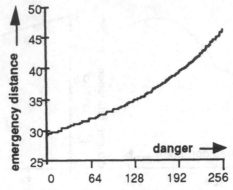

Fig. 18. Emergency distance with x = 0.5 and y = 0

The graph for danger in figure 12 shows the same behaviour as the graph for risk shown in figure 6. This is due to the normalised definition of the linguistic variables. The value of the variable x is varied. The brake retardations a_1 and a_2 are decreased by the factor danger_influence shown in figures 13, 14, and 15. This results in an increase of the stopping distances of cars C2 and C1 and the emergency distance as shown in figures 16, 17, and 18.

All results show that the emergency distance compared to the conventional distance controller is increased as soon as danger or risk are greater than zero. That means that the quality of the control is positively influenced resulting in higher safety.

7 Summary

For the first time a fuzzy model for driver risk and environment danger has been included into the distance controller positively influencing the quality of the control. The consideration of psychological and physical factors using fuzzy logic has been shown to be feasible. The different influences of the variations of the brake retardations a_1 and a_2 and the reaction time r can be adapted choosing appropriate values for x and y (see equation 2). Finally the behaviour of the fuzzy distance controller has been demonstrated by simulation.

It could be shown that the influence of the fuzzy controller behaviour depends more on the rules than on the membership function definitions. This confirms with the statements from [Gupta 1988], [Gariglio 1991] and [Hetzheim 1991], concluding that the choice of the shape of the membership function has small or no influence on the controller's behaviour.

8 Acknowledgements

A major portion of this paper was written when the author prepared her Diploma Thesis in the Daimler-Benz Research Institute (Forschung Systemtechnik) in Berlin.

Special thanks go to Dr. Wolfgang Merker for the support provided to me. The valuable hints of the unknown referees are gratefully acknowledged.

9 References

Burckhardt, M. (1976): Geschwindigkeitsbegrenzung, Sicherheit und Straßenkapazität. Automobiltechnische Zeitung 78 1/2, 23 -27.

Färber, B. (1986): Abstandswahrnehmung und Bremsverhalten von Kraftfahrern im fließenden Verkehr. Zeitschrift für Verkehrssicherheit 32 1, 9 - 13.

Froitzheim, U.F. (1992): Perfekte Symbiose. Wirtschafts Woche Nr 10, 104 - 106.

Gariglio, D. (1991): Fuzzy in der Praxis. Worauf es bei der Entscheidung für die Fuzzy-Technologie ankommt. Elektronik 20, 63 - 75.

Gupta, M.M., Yamakawa, T. (1988): Fuzzy Computing Theory Hardware and Applications. Prentice-Hall Internatinal, London.

Herberg, K.-W. (1985): Alter, Ermüdung und Medikamente als Risikofaktoren. Verkehrssicherheit und Wirksystem Fahrer – Fahrzeug – Umwelt. Verlag TÜV Rheinland, 141 - 175.

Hetzheim, H., Hommel, G. (1991): Fuzzy Logic für die Automatisierungstechnik? atp – Automatisierungstechnische Praxis 33, 10, 504 - 510.

Hör, E. (1983): Die Bedeutung des Faktors Reaktionsdauer für die polizeiliche Verkehrssicherheitsarbeit mit Problemen im Zusammenhang mit der Geschwindigkeitsüberwachung. Verkehrsunfall und Fahrzeugtechnik. 21, 6, 158-166.

Leutzebach, W., Steierwald, G. (1979): Expected improvements in the traffic flow and accident figures by means of autonomous collision avoidance systems. Int. Symp. on Traffic and Transport Technol.; Proc.; Hamburg.

Müller, W. (1991): Alles im Blick. Wirtschafts Woche 38, 76 - 81.

Näätänen, R., Summala, H. (1976): Road-user Behaviour and Traffic Accidents. North-Holland Publishing Company, Amsterdam.

Nicklisch, F., Löffelholz, H. (1989): Fahrzeugtechnische Hilfen zur Einhaltung von Sicherheitsabständen, Ein Sachstandsbericht., Bundesanstalt für Straßenwesen, Bereich Unfallforschung, Bergisch Gladbach.

Palm, R., Hellendoorn, H. (1991): Fuzzy Control. KI 4, 18 - 23.

Sömen, H. (1984): Risikoerleben. Bundesanstalt für Straßenwesen, Bergisch Gladbach.

Togai InfraLogic, Inc. (1990): User's Manual – TILShell, Irvine.

Weidmann, A. (1992): Entwurf eines an Fahrer und Umwelt angepaßten Abstandsreglers mit Fuzzy Logic. Diplomarbeit an der TU Berlin, Fachbereich Informatik.

Wilde, G.J.S. (1978): Theorie der Risikokompensation der Unfallverhütung und praktische Folgerungen für die Unfallverhütung. Hefte zur Unfallheilkunde, 130.

Wocher, B., Nier, J. (1971): Automatische Abstandshaltung zwischen Kraftfahrzeugen als Teilproblem der automatischen Verkehrssteuerung. Bosch Techn. Ber. 3, H. 6, 250 - 255.

Zimmermann, H.-J. (1991): Fuzzy Set Theory – and Its Application. Second, Revised Edition, Kluwer Academic Publishers, Boston/Dordrecht/London.

Workshop Description

Karl Menger,
Fuzzy Logic and Artificial Intelligence –
An Experiment in Reflection

Rainer Born

Inst. f. Phil. and Philosophy of Science
Johannes Kepler University, A-4040 Linz, Austria
tel.: ++43 (0)732 2468-496 (or 499)
E-mail K270190@ALIJKU11 (bitnet) or K270190@EDVZ.UNI-Linz.AC.AT (internet)

Description. The aim of this workshop is an unorthodox "investigation" of the logical and philosophical foundations and of the historical developments concerning the relationship between FL and AI. Special emphasis will be placed upon the mathematical and philosophical contributions of Karl Menger (and perhaps other members of the Vienna Circle in the thirties of this century). Of special interest is the idea of *concept formation* (due to fuzzy set theory or other approaches – e.g. connectionist ones) and the way in which our so-called outer world is grasped or represented by those (perhaps fuzzy) concepts which surely provide the basis for a sound information processing by human beings crediting both scientific and common-sense approaches to knowledge acquisition in general.

Workshop for Doctoral Students in Fuzzy-Based Systems

Johannes Gamper[1] and Bernhard Moser[2]

[1] Institute f. Medical Computer Sciences, Vienna University
Waehringer Guertel 18-20, A-1090 Vienna, Austria
Tel: (+43)-1-40400-3494 Fax: (+43)-1-4052988
email: gamper@wiimc12.imc.univie.ac.at

[2] Institute f. Computer Sciences, Salzburg University
Hellbrunnerstr. 34, A-5020 Salzburg, Austria
Tel: (+43)-662-8044-5325 Fax: (+43)-662-8044-5010
email: 1moser@edvz.uni-salzburg.ada.at

Description. This one-day-workshop held in conjunction with the FLAI'93 conference is intended to enhance the personal and professional development of Ph.D.-level students who are working in the field of fuzzy-based systems. It will enable them to meet, to dicuss their research, and to develop their working skills. It also serves as an introduction to the European Fuzzy-Mail-List established by Wolfgang Slany. This network enables participants to exchange information, give help and support, and to give wider access to resources and methods. Although the network is set up and based in Europe, it is connected to the fuzzy-related list of the North American Fuzzy Information Processing Society, and will be connected to other fuzzy related electronic media in the future.

Attendants of the workshop are asked to either send a one-page abstract of their thesis research or a technical position statement of particular interest to Ph.D. students in fuzzy-based systems. The abstracts will be distributed to the workshop participants and will serve as the basis for the technical discussion during the workshop.

Discussions and sessions will focus on both technical aspects like "How to draft a Ph.D. thesis" and non-technical issues concerning problems of particular interest.

The general form of the workshop will be interactive, focusing on active work, both in small groups and in the plenum. The results will be collected in a report covering the technical as well as the non-technical results of the day. Parts of the report will be written by the participants during the workshop.

Workshop Description

Industrial and Commercial Applications of Fuzzy Logic

Andreas Geyer-Schulz[1] and Peter Kotauczek[2]

[1] Currently: Department of Business Administration / Management Information Systems, University of Augsburg, Memmingerstraße 18, D-8900 Augsburg, Germany.

On leave: Department of Applied Computer Science, Institute of Information Processing and Information Economics, Vienna University of Economics and Business Administration, Augasse 2–6, A-1090 Vienna, Austria

[2] BEKO Gmbh, Weißgerberlände 38, A-1030 Vienna, Austria

Description. The FLAI'93 workshop "Industrial and Commercial Applications of Fuzzy Logic" will provide an informal forum to conference participants to address key issues in industrial and commercial applications of fuzzy logic. The objective of the workshop is to discuss emerging applications of fuzzy logic based technology, to start a dialogue on exploring new ways of applying fuzzy logic and ways to increase the role of fuzzy logic in engineering and decision-making disciplines.

Topics of the workshop will include reports on work in progress and on the current state of the art from the following fields:

- automation and industrial control,
- ecology,
- decision support,
- pattern recognition,
- image generation,
- ...

Fuzzy logic based technology is already impacting on the way products are developed in the automation and industrial control market and in the consumer products market causing changes in the product life-cycle, the development costs and time to market. What do these changes mean to companies in these markets? And, can we expect this impact to spread into other markets, like for example the financial services sector?

The key to understanding the impact of fuzzy logic based technologies is understanding that fuzzy logic is a true paradigm shift. The shift is bound to impact on the way how new products are developed and how long it takes to market them. A considerable reduction of development costs is to be expected for some products, for example household appliances. We may see new markets emerging as fuzzy logic based technologies become commercially available. Commercial aspects, technology assessment as well as comparative studies on these effects are part of the workshop.

The last part is devoted to practical problems. Problem areas, where fuzzy technology is expected (or known) to be less appropriate and successful, shall be identified and discussed with the goal of sharing experiences and finding ways to overcome such obstacles by, for example, hybrid techniques. Finally, especially for companies entering the field, development methodology and phase models for fuzzy system development are of considerable interest and can lower the entry barrier and the entry cost.

Workshop Description

Fuzzy Scheduling Systems

Roger Kerr

Christian Doppler Laboratory for Expert Systems
Technical University of Vienna
Paniglgasse 16, A-1040 Vienna, Austria

and

School of Mechanical and Manufacturing Engineering
The University of New South Wales
P.O. BOX 1, Kensington, New South Wales, Australia 2033

Description. Scheduling is an activity which is usually based on uncertain time estimates and loosely specified preferences and constraints which can often be relaxed. An increasing number of researchers have recently started to explore the potential uses of fuzzy set theory, fuzzy logic and fuzzy arithmetic to assist in a more realistic representation of the uncertainty and vagueness typically inherent in schedules. This workshop is intended to provide a forum for those interested in this field to meet for presentation and discussion of recent research, to raise possible research or philosophical issues, and to report on applications.

Themes:

- fuzzy scheduling heuristics
- fuzzy preference constraint representation and relaxation
- temporal constraint representation and propagation using fuzzy arithmetic
- fuzzy knowledge elicitation: mapping linguistic scheduling variables and preferences on to fuzzy sets
- methods of comparative evaluation of crisp and fuzzy schedules
- philosophical implications
- case studies

Springer-Verlag
and the Environment

We at Springer-Verlag firmly believe that an international science publisher has a special obligation to the environment, and our corporate policies consistently reflect this conviction.

We also expect our business partners – paper mills, printers, packaging manufacturers, etc. – to commit themselves to using environmentally friendly materials and production processes.

The paper in this book is made from low- or no-chlorine pulp and is acid free, in conformance with international standards for paper permanency.

Lecture Notes in Artificial Intelligence (LNAI)

Lecture Notes in Computer Science